THE

FIRST

HUMAN

 DOUBLEDAY

NEW YORK   LONDON   TORONTO   SYDNEY   AUCKLAND

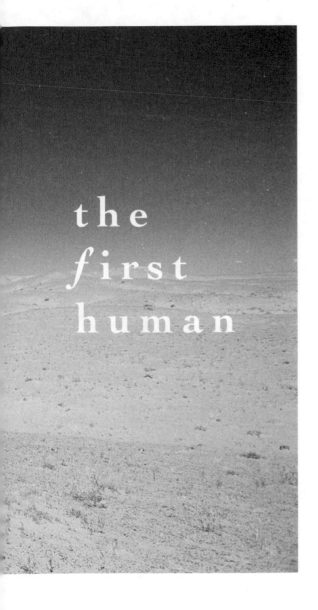

the
first
human

THE

RACE

TO

DISCOVER

OUR

EARLIEST

ANCESTORS

ANN
GIBBONS

PUBLISHED BY DOUBLEDAY
a division of Random House, Inc.

DOUBLEDAY and the portrayal of an anchor with a dolphin are registered
trademarks of Random House, Inc.

"Every Grain of Sand" by Bob Dylan, © 1981 by Special Rider Music.
All rights reserved. International copyright secured. Reprinted by
permission.

Book design by Maria Carella

Map and time line designed by Jeffrey L. Ward

Photograph of desert landscape courtesy of Photodisc Collection/
Getty Images

Library of Congress Cataloging-in-Publication Data
Gibbons, Ann.
   The first human : the race to discover our earliest ancestors / Ann
Gibbons.—1st ed.
      p. cm.
   Includes bibliographical references and index.
   1. Fossil hominids. 2. Human evolution. I. Title.
   GN282.G52 2006
   599.93'8—dc22   2005053780

ISBN 0-385-51226-0

PRINTED IN THE UNITED STATES OF AMERICA

10 9 8 7 6 5 4 3 2 1

First Edition

FOR BILL

AND OUR DESCENDANTS:

LILY, SOPHIA, AND TOM

It has been said, that the love of the chase is an inherent delight in man—a relic of an instinctive passion.

CHARLES DARWIN
Diary of the Voyage of H.M.S. Beagle

CONTENTS

PART THREE
WISDOM OF THE BONES / 223

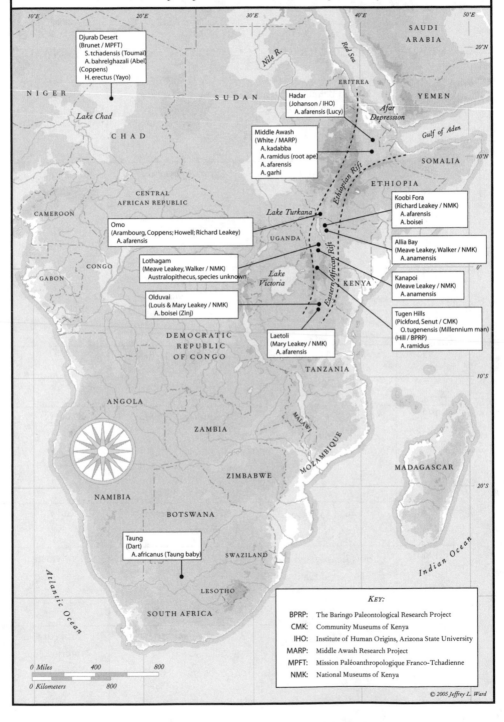

THE CRADLE OF HUMANITY

The African sites where the fossils proposed as the earliest known members of the human family have been discovered in the past century

Djurab Desert
(Brunet / MPFT)
S. tchadensis (Toumaï)
A. bahrelghazali (Abel)
(Coppens)
H. erectus (Yayo)

Hadar
(Johanson / IHO)
A. afarensis (Lucy)

Middle Awash
(White / MARP)
A. kadabba
A. ramidus (root ape)
A. afarensis
A. garhi

Koobi Fora
(Richard Leakey / NMK)
A. afarensis
A. boisei

Omo
(Arambourg, Coppens; Howell; Richard Leakey)
A. afarensis

Allia Bay
(Meave Leakey, Walker / NMK)
A. anamensis

Lothagam
(Meave Leakey, Walker / NMK)
Australopithecus, species unknown

Kanapoi
(Meave Leakey / NMK)
A. anamensis

Olduvai
(Louis & Mary Leakey / NMK)
A. boisei (Zinj)

Tugen Hills
(Pickford, Senut / CMK)
O. tugenensis (Millennium man)
(Hill / BPRP)
A. ramidus

Laetoli
(Mary Leakey / NMK)
A. afarensis

Taung
(Dart)
A. africanus (Taung baby)

KEY:

BPRP: The Baringo Paleontological Research Project
CMK: Community Museums of Kenya
IHO: Institute of Human Origins, Arizona State University
MARP: Middle Awash Research Project
MPFT: Mission Paléoanthropologique Franco–Tchadienne
NMK: National Museums of Kenya

0 Miles   400   800

0 Kilometers   800

© 2005 Jeffrey L. Ward

"FIRST HUMAN" FOSSIL FINDS BY YEAR

YEAR	DISCOVERER	SCIENTIFIC NAME	NICKNAME	
2002	Michel Brunet and MPFT	*Sahelanthropus tchadensis*	Toumaï	
2000	Martin Pickford and Brigitte Senut	*Orrorin tugenensis*	Millennium Ancestor	
1996	Yohannes Haile-Selassie	*Ardipithecus kadabba*		
1995	Michel Brunet and MPFT	*Australopithecus bahrelghazali*	Abel	
1994	Meave Leakey and the Hominid Gang	*Australopithecus anamensis*		
1992	Gen Suwa and Tim White	*Ardipithecus ramidus*	Root Ape	
1974	Donald Johanson	*Australopithecus afarensis*	Lucy	
1960	Louis Leakey and the Hominid Gang	*Homo habilis*	Handyman	
1959	Mary and Louis Leakey	*Australopithecus boisei*	Zinj (or Nutcracker Man)	

AGE	LOCATION	SIGNIFICANCE
6–7 million	Toros-Menalla, Chad	Proposed as earliest known hominid. Teeth and skull show "human" features, and possibly signs of upright walking. Cradle of humanity may have been in central Africa.
6 million	Tugen Hills, Kenya	Proposed as earliest known hominid. Thighbone shows evidence of upright walking.
5.8 million	W. Margin, Middle Awash, Ethiopia	Toe bone shows possible evidence of upright walking. Teeth show some human traits not found in apes.
3.5 million	Koro Toro, Chad	First fossil of early hominid from central Africa, which proves hominids already ranged beyond eastern Africa. Resembled Lucy's species (and may be *A. afarensis*).
3.9–4.1 million	Kanapoi, Kenya	Shinbone shows upright walking was fully assembled in hominids by 4.1 million years ago.
4.4 million	Aramis, Ethiopia	Proposed as earliest known hominid. Primitive baby molar and possible evidence of upright walking. Earliest known skeleton under analysis as Rosetta stone of early bipedalism.
3.1 million	Hadar, Ethiopia	Earliest member of human family for 20 years. Small brain, yet walked upright.
1.8 million	Olduvai, Tanzania	First direct ancestor of humans found in eastern Africa. Showed increase in brain size and smaller chewing molars than australopithecines.
1.8 million	Olduvai, Tanzania	First member of the human family from eastern Africa.

(continued)

YEAR	DISCOVERER	SCIENTIFIC NAME	NICKNAME	
1934	G. Edward Lewis	*Ramapithecus brevirostris*	Rama's Ape	
1929	Davidson Black	*Sinanthropus pekinensis*	Peking Man	
1924	Raymond Dart	*Australopithecus africanus*	Taung Baby	
1891	Eugène Dubois	*Homo erectus*	Java Man	

AGE	LOCATION	SIGNIFICANCE
8–13 million	Siwalik Hills, Pakistan	Long thought to be the earliest ancestor of humans; showed that teeth and jaw fragments alone were unreliable to prove hominid status.
0.2–0.5 million	Zhoukoudian, Beijing, China	Skull resembled Java man's; later both would be named *Homo erectus*.
1–2 million	Taung, South Africa	Relatively small brain with big forebrain, walked upright.
0.8–1.2 million	Trinil, Java	First fossil of early hominid. Suggested Asian origin for humankind.

The Human Family

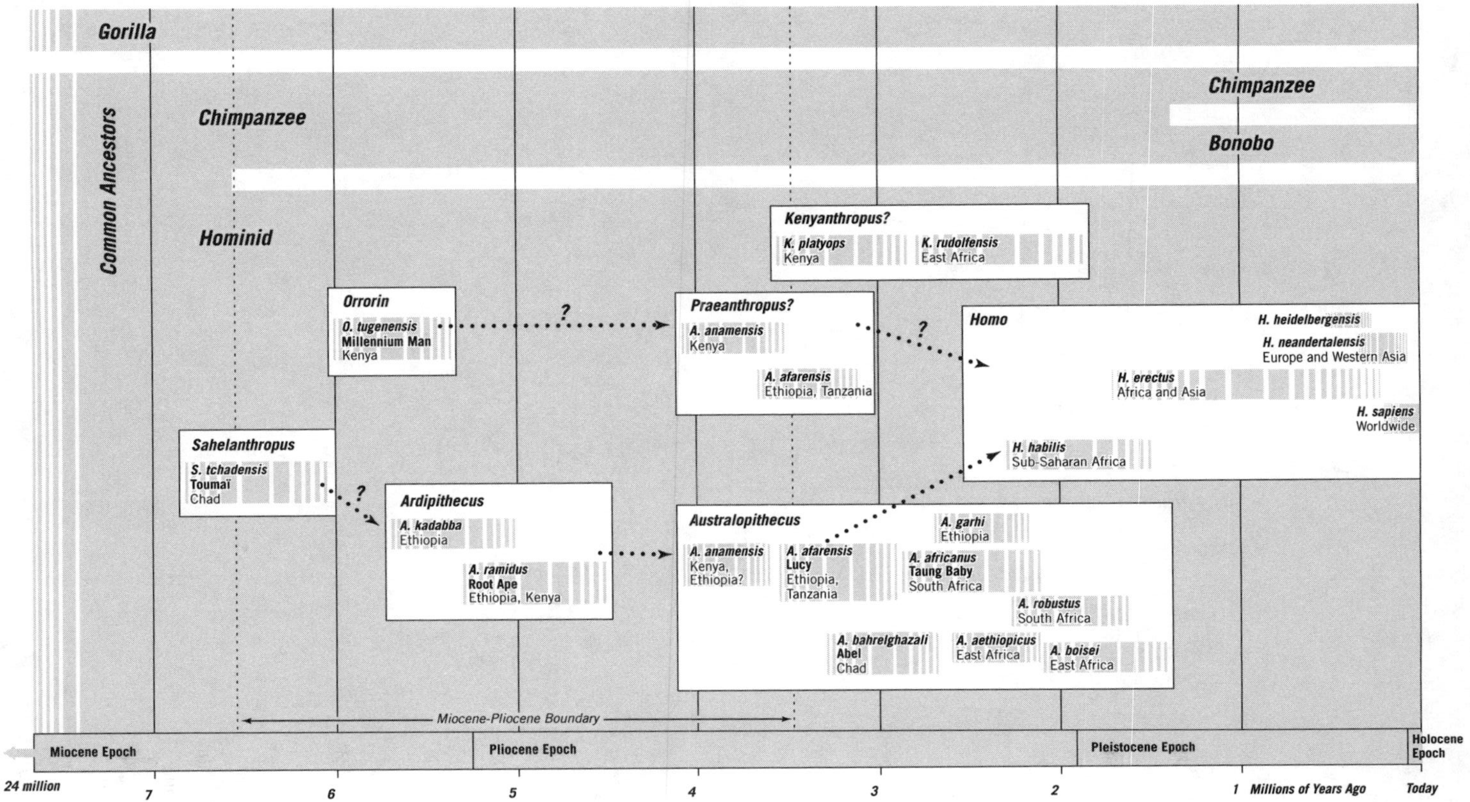

THE FOSSIL HUNTERS

Berhane Asfaw

Ethiopian biological anthropologist; coleader of the Middle Awash Research Group; director of the Rift Valley Research Services in Ethiopia.

Alain Beauvilain

French geographer who coordinated logistics and surveys for the Mission Paléoanthropologique Franco-Tchadienne (MPFT) in Chad from 1994 until late 2002; lecturer at the University of Paris X–Nanterre.

Michel Brunet

French paleontologist; leader of the Mission Paléoanthropologique Franco-Tchadienne (MPFT) that discovered Toumaï and Abel; professor at the University of Poitiers in France.

Desmond Clark

British-born archaeologist who was head of the first Berkeley team to explore the Middle Awash; professor emeritus at Berkeley when he died in February 2002.

Yves Coppens

French paleoanthropologist who was a codiscoverer of fossils of Lucy's species in Ethiopia; collaborator with Michel Brunet and Martin Pickford and Brigitte Senut; professor at the Collège de France in Paris.

Raymond Dart
Australian-born anatomist who discovered the first fossil of a human ancestor from Africa, from Taung, South Africa, in 1925. He died in 1988.

Ahounta Djimdoumalbaye
Chadian student at the University of N'Djamena who discovered the skull of Toumaï on July 19, 2001; member of the Mission Paléoanthropologique Franco-Tchadienne (MPFT).

Eugène Dubois
Dutch anatomist and paleontologist who discovered the first fossils of a hominid, Java man, in 1890 in Java, Indonesia. He died in 1940 in the Netherlands.

Eustace Gitonga
Kenyan artist; director of the Community Museums of Kenya, the organization that obtained permits for Martin Pickford and Brigitte Senut to do research in the Tugen Hills.

Yohannes Haile-Selassie
Ethiopian paleoanthropologist who discovered the partial skeleton of *Ardipithecus ramidus* and fossils of *Ardipithecus kadabba* in the Middle Awash of Ethiopia; member of the Middle Awash Research Group; curator of physical anthropology at the Cleveland Museum of Natural History.

Andrew Hill
British-born geologist; director of the Baringo Paleontological Research Project in the Tugen Hills of Kenya; chairman of the anthropology department at Yale University.

Clark Howell
American paleoanthropologist; coleader of the Omo Expedition to Ethiopia in 1966; professor emeritus at the University of California, Berkeley.

Donald Johanson

American paleoanthropologist who discovered Lucy's skeleton in 1974. He is the director of the Institute of Human Origins at Arizona State University in Tempe.

Jon Kalb

American geologist; member of the first French-American team to explore Hadar, in 1971; leader of the first team to find fossils of a hominid in the Middle Awash; research fellow at the University of Texas at Austin.

Louis Leakey

Pioneer British, Kenyan-born anthropologist and paleontologist who established East Africa as a critical place for finding human ancestors; codiscoverer of Zinj in 1959; director of the Coryndon Museum (now the National Museums of Kenya). He died in 1972.

Mary Leakey

British-born illustrator and self-trained paleontologist who found the skull of Zinj in 1959 at Olduvai, Tanzania. She died in 1996.

Meave Leakey

Welsh-born zoologist whose team found fossils of *Australopithecus anamensis* at Kanapoi; retired director of paleontology at the National Museums of Kenya.

Richard Leakey

Kenyan paleontologist; former director of the National Museums of Kenya; visiting professor of anthropology at Stony Brook University in New York.

Bryan Patterson

British-born paleontologist who discovered fossils of early hominids in the mid-1960s at Kanapoi and Lothagam; professor at Harvard University. He died in 1979.

Martin Pickford

British-born geologist who discovered fossils of Millennium Man in the Tugen Hills; coleader of the Kenya Paleontology Expedition; geologist at the Collège de France in Paris.

David Pilbeam

British-born paleoanthropologist whose team discovered fossils of *Ramapithecus* in Pakistan; professor of anthropology at Harvard University.

Vincent Sarich

American molecular anthropologist; professor emeritus of anthropology at the University of California, Berkeley.

Brigitte Senut

French paleontologist who discovered fossils of Millennium Man in the Tugen Hills; coleader of the Kenya Paleontology Expedition; professor at the National Museum of Natural History in Paris.

Elwyn Simons

American primate biologist who proposed *Ramapithecus* as an early hominid in the 1960s; professor of biological anthropology, anatomy, and zoology at Duke University in Durham, North Carolina.

Gen Suwa

Japanese paleoanthropologist who discovered the first fossil of *Ardipithecus ramidus*, in Ethiopia; associate professor at the University Museum, the University of Tokyo, in Japan.

Maurice Taieb

French geologist who discovered the fossil beds at Hadar where Lucy was found and in the Middle Awash where *Ardipithecus* was found; director of research emeritus for CNRS-CEREGE laboratory in Aix-en-Provence, France.

Alan Walker

British-born paleontologist who found fossils of *Australopithecus anamensis* at Allia Bay, Kenya; professor of anthropology at Pennsylvania State University.

Tim White

American paleoanthropologist; coleader of the Middle Awash Research Group; professor at the University of California, Berkeley.

Giday WoldeGabriel

Ethiopian geologist; coleader of the Middle Awash Research Group; geologist at the Los Alamos National Laboratory in New Mexico.

THE

FIRST

HUMAN

On a summer day in July 1995, Michel Brunet had a premonition that he was on the verge of a major discovery. Brunet, who was a little-known fifty-five-year-old French paleontologist, had traveled to the National Museum of Ethiopia in Addis Ababa to seek wisdom from the ancient bones stored in its vaults. But access was not assured, so he came with an offering of his own: a 3.5-million-year-old jawbone that he had found among the shifting sands of the Djurab Desert of Chad. He hoped to compare it with the pantheon of famous fossils locked away in the museum, including the earliest members of the human family then known.

There to greet him was Tim White, a forty-four-year-old paleoanthropologist from Berkeley who was emerging as the most successful fossil hunter of his generation. With his biting wit and notorious impatience with those he does not respect, White can be daunting to those seeking access to fossils his team has found. Indeed, his team has turned away researchers who flew to Ethiopia to see fossils still under study. But White soon recognized a fellow traveler in Brunet: both share a fundamental passion for fossils. Both are tireless in the field, returning every year. And both have risked their lives to find fossils: White has suffered from malaria, giardia, dysentery, hepatitis, and pneumonia and Brunet struggles with a heart condition. Brunet was no "salon anthropologist" who wanted a preview of fossils that White and his colleagues were still analyzing.

Before long, White and two of his former graduate students were opening the museum safes for Brunet to show him fossils of the same age as his jawbone so he could sort out its identity. As they studied the bones and compared notes about the fossil beds where they had worked, Brunet had a revelation: his team had found the same kinds of animal bones at sites in Chad that White and his colleagues had found with their fossils of early human ancestors in Ethiopia.

This was encouraging to Brunet, because White and his colleagues had discovered the earliest member of the human family known at the time—a creature the size of a chimpanzee named *Ardipithecus ramidus* that lived in the woods of the Ethiopian Rift Valley 4.4 million years ago. White's team had found particular species of extinct pigs, colobine monkeys, and carnivores nearby. Brunet's team also had found these same types of animals in sediments of similar age—and he began to wonder if he had come across an early habitat for humanity in Chad as well. Perhaps he was hot on the trail of some of the earliest members of the human family?

But when Brunet mentioned that he had also found some extinct gerbils, White shook his head. "Those little rodents live out in the dry places," he said. "Early hominids are in the woods." He told Brunet he would not find any hominids there.

Brunet told White that maybe he was right. But he also knew about even older sediments in Chad—fossil beds where the animal bones were at least 6 million years old, when the habitat may have been wooded. Although his team had yet to explore those sandy beds, he knew they were older than the fossil sites in the Middle Awash where White's student Yohannes Haile-Selassie had recently begun research. Brunet had an inkling that his jawbone was only a prelude of even older discoveries to come from beneath the dunes of the Djurab Desert, which he had just begun to explore.

Brunet made a bold prediction on that day in 1995: he bet White that, gerbils notwithstanding, he would find the oldest hominid, the elusive missing ancestor. The underdog Frenchman would beat the better-financed and better-known American. "I am working in older sediments," he said, half jokingly. "I will win."

INTRODUCTION

As the sun rose over an open-air camp in the Djurab Desert of Chad, Michel Brunet sat up on his cot and saw nomads appear over the dunes with their camels, almost like an apparition in the early-morning light. At first he tensed, wondering if these Arab men and women were members of the warlike northern tribe that had waged war with southern tribes for thirty years on the desolate land, littering it with mines. But when the men in flowing blue-and-white robes and turbans smiled and the women offered him camel's milk and tea, he realized they were Gorane camel herders, who roamed the region in search of water. He relaxed as his team members conversed with them. As Brunet prepared to leave, the nomads offered a customary blessing to him in their language, asking Allah to protect him and give him happiness.

Later that morning, Brunet felt a bit like a nomad himself as he headed out into the sand dunes. He must have been an odd sight to the nomads who had come to watch him work. Now in his mid-sixties, he can resemble a bearded version of the actor Anthony Hopkins, with silver hair swept back to reveal a high forehead and startling blue eyes. But on the morning of January 23, 1995, Brunet wrapped his head in a cloth and donned a ski mask as he prepared for another day's work in the Djurab, where sand blows relentlessly into a person's eyes, ears, nose, and mouth. Temperatures can get so hot that plastic drink containers left in the shade of cars and tents—there is no other shade—can spontaneously explode.

As Brunet joined the other members of his party, they spread out across the area, each one walking slowly, stooped at the waist so they would not miss anything as they scoured the ground. They were sweeping the desert floor for bones, passing back and forth over the same terrain so they would not overlook even the smallest fossils. They were careful not to touch anything metal, in case it was one of the mines left by the northern rebels— deadly reminders of the civil war waged with government forces in this desolate region. It was tedious work that would have to stop as the sun rose higher and temperatures soared. Brunet was trying to keep focused on the rubble in front of him when he spied a bone protruding from the sand. He let out a cry. But it was only the fossil of a pig.

Then the Chadian driver, Mamelbaye Tomalta, called Brunet to come over to him. He had found a jaw with its teeth dug into the ground. Brunet would never forget what he saw as he brushed off the sand. It looked like the jaw of an ancient ape, but the shape of its teeth startled him. They had a closer resemblance to those of a human. He quickly recognized that he was looking at the jawbone of an early human ancestor who had lived on the ancient shore of Lake Chad about 3.5 million years ago.

Later that night, Brunet could not sleep. As he lay awake on his cot, he remembered the nomads' blessing and wondered if this jawbone would indeed bring him happiness. He got up twice just to shine his flashlight on the jawbone, while members of his team slept beside him in a tent. He wanted to make sure that he wasn't dreaming—that the jawbone was real. The only shadow that clouded the moment was that he could not show it to his long-time collaborator, geologist Abel Brillanceau, who had died six years earlier of drug-resistant malaria when he and Brunet were searching for fossils in the woods of Cameroon. Brunet vowed that night to name the jawbone Abel. A few days later, when he got to a telephone in N'Djamena, he called another longtime friend and colleague, paleoanthropologist David Pilbeam of Harvard University, who had been part of the mission in Cameroon. It was early in the morning in Cambridge, Massachusetts, and Brunet woke up Pilbeam. He said: "David, I've got it." Pilbeam knew immediately what he meant, and was deeply pleased for Brunet. If anyone deserved a discovery like this, it was Brunet.

The jawbone was a long-sought prize for Brunet, who had never before found a fossil of an early human ancestor. He certainly had tried—by the time he found the jawbone in 1995, Brunet had made his name as a paleontologist's paleontologist, and was respected for his skill in finding animal fossils in some of the world's most remote and hostile sites. His adventures in the field were legendary: he had been strafed by a fighter jet in Afghanistan, arrested in Iraq, threatened at gunpoint in Chad. But still, even when he was down on his luck and without funding for his fieldwork, he persisted. Year after year, he left his lab at the University of Poitiers, in central France, to return to the field, even scouting new sites in the Djurab Desert in a rented four-wheel-drive car with barely enough water to brush his teeth. He persevered, and it paid off—his team found thousands of fossils of extinct monkeys, elephants, giraffes, rhinoceroses, hippopotamuses, and pigs. But one type of mammal eluded him: a hominid. Until that day in January, Brunet had never even held an actual fossil of an early hominid, only casts of fossils found by others. So when he finally cradled the hominid jawbone in his hands, it was a life-transforming moment. Nineteen years of searching for the fossil of a hominid had been "a long time in the life of a small bipedal ape," Brunet would say, referring to himself.

What is it that makes a piece of gray jawbone with rotting teeth so tantalizing? Why was it worth risking his life to find? Brunet throws up his hands in a characteristic French way, exhaling, "Pfew!" He is an intense man with a short fuse who describes himself as "crazy, French, poor, a Socialist"—albeit a Socialist who drives a Mercedes. As if it is obvious, he declares that he wants to know where humans come from. He has been obsessed with the question ever since reading Darwin's *The Descent of Man,* written in 1871. Darwin proposed that humans originated in Africa because African chimpanzees and gorillas are the great apes most closely related to humans, and that we are all members of the primate order. Ever since, explorers have been searching for this missing link, a name that comes from the old idea of the Great Chain of Being, in which creatures of the earth are connected one to the other, from lowest to highest. Scientists have been looking for fossils that would show how humans fit in nature and their place in the animal kingdom.

Starting with Dutch anatomist Eugène Dubois's discovery of Java man in Indonesia in 1891, many fossils have been proposed as the missing link, only to be bumped from that spot when an even older and more primitive fossil was found.

The fossil that reigned the longest as the earliest human ancestor was Lucy, a female the size of a chimpanzee whose remarkable partial skeleton was discovered in Ethiopia in 1974 by a young American paleoanthropologist, Donald Johanson. Lucy was a member of the species *Australopithecus afarensis*, which lived in the Great Rift Valley of eastern Africa 3 million to 3.6 million years ago. For twenty years, textbooks put Lucy's species as the ancestor of all humankind, giving rise to humans who came later and some extinct lineages of ape-men who lived in Africa, as well. It was a neat line of descent, pleasing in its orderly unfolding of one species into another.

By the mid-1990s, when Brunet began scouting for sites in Chad, there were unmistakable clues that this view was too simplistic. The human story was beginning to look as complicated as a Tolstoy novel, with new characters appearing unexpectedly as the book of life unfolded. Although most researchers thought that Lucy's species begot the lineage that led to modern humans, there were suggestions that she was not the only type of early human on the planet 3 million to 4 million years ago. New fossils added new branches to the human family's ancestral tree. Some branches led nowhere and represented extinct lineages. Other lineages existed simultaneously from 1 million and 3 million years ago, igniting debate about which hominids were on the one line of descent down to modern humans. It had also been obvious for some time that there was not just one missing link between humans and the ancestor of apes over the course of the human story—there were many missing links on the one true line to humans over millions of years. And they were not perfect intermediates that looked half ape and half human. The term "missing link" fell into disfavor.

At the same time as new fossils were making their appearance, there was a revolution in the field of molecular biology. The molecular evolutionists had suggested in the 1960s that the earliest players in the human story had yet to be found. Most anthropologists had not believed those findings. By the mid-1990s, the molecular evidence was so strong that it was clear that the first chapter—the genesis of humankind—was missing entirely. Biochemists had identified chimpanzees as the closest living relative of humans by comparing the DNA of humans and other apes. When they lined up the same stretches of DNA from humans and chimpanzees, they consistently found too many differences—or mutations—to have accumulated in the time since Lucy's species arose, about 3.8 million years ago. Since mutations accumulate over long stretches of DNA at a relatively steady rate over millions of years, geneticists can count the differences in the DNA and use them like a clock to date roughly when one species splits from another species. The molecular clock, which is set using dates from the fossil record, has consistently timed the split of humans from the ancestor of chimpanzees earlier—to sometime between 5 million and 7 million years ago. Yet fossil hunters had found only a few teeth and scraps of poorly dated bone older than Lucy. The greatest unsolved problem in the field of human origins was, What came before Lucy? Who was the first member of the human family? And where was the even older ape that was the last ancestor humans shared with chimpanzees before they parted company and went down their separate evolutionary paths?

It was this mystery that drove Brunet and a handful of other fossil hunters to new fossil sites in Africa in the 1980s and 1990s. Brunet and Pilbeam went west and others, including Tim White, went east. They all felt poised to find the earliest members of the human family—or at least our oldest ancestors' closely related contemporaries, since the chances of finding the actual individuals who were our direct ancestors are vanishingly low. With the aid of modern genetics to help them decide what time horizons to explore and new technical methods to date fossils, they knew they were closing in on fossils old enough to have lived soon after the ancestor of humans split from the ancestor of the African apes. The fossil trail led most of these fossil hunters to eastern Africa, long considered the cradle of humanity, because the fossil record outside eastern Africa and South Africa had failed to break the 2-million-year barrier. Brunet's strategy to search in western and central

Africa was a gamble, since no human ancestors older than 1 million years had been found in Chad.

So when news spread that Brunet had encountered the jawbone of Abel on that January morning in 1995, the discovery reverberated through the paleoanthropological world. It opened a third window into the earliest stages of human evolution, adding central Africa to the view that had encompassed only eastern Africa and South Africa. The discovery made Brunet a bit of a celebrity in France, where tracing human origins is a national pastime. At an age of 3 million to 3.5 million years, the jawbone was the oldest fossil of a hominid found outside of eastern or southern Africa. This meant that the earliest hominid had to be much older—and might have arisen outside eastern Africa.

For a field-worker little known outside of paleontological circles, the attention was alluring. Brunet clearly enjoyed the hominid aura, but he knew that his brief flirtation with fame was nothing like the immortality that would befall the discoverer of the first hominid. No Nobel Prizes are awarded to anthropologists or paleontologists, but the discoverer of the earliest human ancestor would surely become as famous as the fossil itself. In the field of human origins research, the names of discoverers are forever linked with the famous fossils they have found: Eugène Dubois and Java man, Raymond Dart and the Taung baby, Louis and Mary Leakey and Zinj, Donald Johanson and Lucy. Brunet was wary, however, of the insidious influence of fame. He observed years later that many of the paleoanthropologists who have found the most famous fossils have done less science over the years. Some were demoralized by criticism of their fossils, which they were as attached to as their own progeny. Others, who found quick fame, spent more and more time on the lecture circuit and in front of television cameras, making documentaries of their discoveries, publishing their memoirs, or raising funds for their teams. In 1995, Brunet was middle-aged, had a heart condition, and knew he had much work still to do, and so he made a conscious decision to keep his sights on the science. But he also was pragmatic—he capitalized on his newfound notoriety and used it to convince French politicians to build a museum for fossils in

N'Djamena, the capital of Chad, and to buy his team a small, lightweight plane for field missions in Chad.

For his search for the first hominid had only just begun. Even as he toured the labs of colleagues with a cast of his jawbone, he already had an inkling that it was a forerunner of better fossils to come from beneath the sand dunes in Chad. He knew he was well positioned to find something even older, and closer to the origins of the human lineage, because he had found fossils of other mammals that lived more than 6 million years ago in the Djurab Desert. He could not date these fossils directly, but he knew they were species of animals that went extinct more than 6 million years ago, based on the discovery of the same species at other reliably dated sites in Africa. These mammals were beacons that guided Brunet—pointing the way to the fossil beds that opened a window into the Late Miocene, an epoch 5.3 million to 11 million years ago. This was well before Lucy lived, and it was a mysterious era in which practically no fossil apes had been found in Africa. After the discovery of the Abel jawbone, Brunet focused his sights on these fossil beds, starting in 1997. His team spent week after week sweeping the surface of the dunes and sifting fossils ranging from rodent teeth the size of salt crystals to the stretched-out snouts of hippopotamuses. Before long, they came upon an ancient crossroads for many different kinds of animals drawn to the gallery of trees along the shore of the ancient Lake Chad.

Conditions were daunting, even by Brunet's standards. In some years, windstorms buried their tents, trapping them inside for several days, where they survived on pasta and tuna, rice and sardines. When they finally were able to venture outside, they had to dig out of the sand, as if it were snow, and keep within sight of each other so they would not become disoriented in the windstorms.

But the wind was also their ally. Every year, windstorms erode more than an inch of sandstone, sending dunes rippling slowly across the flat desert like waves on a sea and exposing fossils that have been buried for millions of years. This bounty of fossils left in their wake gave Brunet the confidence to renew his quest. He wanted nothing less than to find the earliest human ancestor. And he knew he had no time to waste—two other teams were already ahead of him, including White's. They were well on their way, following the trail of the most ancient ones—the ancestors who lived more than 4 million years ago. Who would succeed?

∽∽∽

In January 1995, the same month that Brunet discovered his jawbone in Chad, Tim White was twenty-five hundred kilometers to the east, sweating on a treeless slope in the badlands of western Ethiopia, poised to remove a skeleton. Day after day, White rose before dawn, ate breakfast, and drove from his camp with his team of international scientists to the famed fossil beds of the Middle Awash, about seventy-five kilometers south of the place where Lucy had been found twenty-one years earlier. Armed with hypodermic syringes, White slowly injected drops of glue into a crumbly fossil. He waited for the bone to set and then, with the help of his colleagues, painstakingly extracted the block of volcanic sediment encasing the fossil, so it could be driven by jeep at the end of the season to a lab at the National Museum of Ethiopia in Addis Ababa for final extraction. Sometimes it took him three days in the museum lab to remove a single bone from the basaltic sediments. In this case, the effort was worth it: this partial skeleton was being called the discovery of a decade.

"Fossils are not found according to *National Geographic* mythology— tanned camel riders stooping occasionally to pick up a nice hominid fossil," says White. He reminds one less of Stanley and Livingstone searching for the source of the Nile and more of a rebel with a cause, seeking meaning from the fossils and intolerant of romantic stereotypes of Indiana Jones–style fossil hunters who get lucky and stumble over fossils. With his complex character and dark humor he could have sprung from a Hemingway novel. Not married until he was fifty-three, he has the single-mindedness of someone who has dedicated his life to his work. He lets nothing—and no one—get in the way of "getting it right." With his wiry frame and focused energy, he is a disciplined, strategic thinker who is particularly good at recognizing the talents (and shortcomings) of his students and team members. He is meticulous and uncompromising when it comes to scientific accuracy. And he doesn't mince words when he disagrees with colleagues about their research or behavior. He once published a list of recommendations to paleoanthropology students that became known as the "Tim Commandments" and included such rules as: "Do not purchase fossils," "Do not bribe officials," "Do not steal another person's site," and "Do not let ambition distort your ethics. If your career goal is to make a fortune, then go to medical school, become a knee surgeon, and practice on suburban soccer players."

He is particularly scathing in his criticism of "careerist" anthropologists who pander to the press and spin "just so" stories about the emergence of humankind, like latter-day Rudyard Kiplings. Yet White is adept at working with the media himself, although he resists the temptation to bring the fossils to life with descriptions of how they might have behaved in their lost worlds until he has analyzed them in detail. When asked on National Public Radio whether Neandertals had speech, for example, he drew a distinct line between science fact and science fiction. "Jean Auel has sold an awful lot of books with stories about that," he said, referring to the best-selling author of *Clan of the Cave Bear* and other Neandertal fiction. He reflexively tries to puncture any notions of glamour surrounding his quest for human ancestors. As a guest lecturer at a college in Minnesota, his response to an audience member's question about how he felt when he walked where humans first walked was typical: "Hot."

White demands similar discipline from the scholars who are members of the Middle Awash Research Group. The group is an international team that over twenty-five years has grown to include more than fifty specialists, most with Ph.D.s, from sixteen nations who work with White on the fossils, site geology, and archaeology. Officially, there are several coleaders of this team, but unofficially White is the commander—indeed, he is sometimes referred to as "the General," and he even sounds like one when he refers to the European or African "theaters" for finding fossils. He set the tone and style for the group early on, although several of the Ethiopians have emerged in recent years as independent researchers in their own right, and although he insists that every major decision about the project is made as a group. He is even careful about how he poses for photos with his Ethiopian colleagues, so that he does not appear to be in charge. As a rule, details about discoveries are kept quiet until they are ready to be published in scientific journals. This is one team that has not been shadowed by film crews and whose movements have not been shown on the National Geographic or Discovery channels.

White's silence about the fossils his team has found creates a mystique that spurs more interest in the team's discoveries; he is an alpha male in a crowd of paleoanthropologists. His rivals keep an eye on him, both intensely curious about what he is doing and wary that they will be stung by his sharp criticism, often thinly veiled as caustic humor. Though most respect his high standard of scholarship and his capacity for hard work, many resent his refusal to offer previews of his fossils or discuss them until he is ready to publish a detailed description—a process that has left the entire field of paleoanthropology in suspense about the identity of the partial skeleton he has been excavating and analyzing in Ethiopia for more than a decade. One colleague calls the Middle Awash research project the "Manhattan Project of Paleoanthropology." But White will not be pressured or rushed, and he is particularly loath to see researchers speculate about the fossils he has under study until he can publish his analysis and influence the first impressions of the fossils. He figures it is easier to say no to everyone rather than give favored colleagues what he calls a "sneak preview," a form of insider trading of scientific information that White thinks is unfair—unless, like Brunet, they have discovered a new fossil that they need to compare directly so they can identify it. This policy does not win him friends. White obviously doesn't care, and he is even somewhat amused by his bad-boy notoriety, chuckling when he learns that he has been called "He Who Shall Not Be Named"—a reference to the brilliant dark wizard who has fallen from grace in the Harry Potter books. What matters more than his colleagues' good opinion is clearly his own opinion about how well he has analyzed and treated the fossils in his care—and his team's track record of finding a stunning array of new fossils, year in and year out. One lucky discovery doesn't impress him. He cites as models two baseball stars, Barry Bonds (*if* he did not use steroids, White notes) and Babe Ruth: "Bonds and the Babe are the best because they built up a record of success."

White's passion for fossils and exploring the natural world began when he was a child, growing up in the San Bernardino Mountains of California, near Lake Arrowhead. His family's home was perched on a ridge in a place with the evocative name of Sky Forest, built on a crest that the locals called the Rim of the World. From this vantage point, thousands of feet above the San Bernardino Valley, White grew up playing in a national forest, catching snakes and horned toads, and climbing on huge granite boulders. Ancient American Indians had ground acorns in the mortars in these boulders, leaving a trail of obsidian flakes and pottery shards, which White collected.

White was still catching rattlesnakes when he was an undergraduate at the University of California in Riverside—and he would later joke that snake recognition would prove a valuable career asset in the contentious field of paleoanthropology. But his childhood fascination with the ancient Indians of California grew into a desire to study the even deeper past and track the way humans and other animals evolved over millions of years in Africa.

On this January day in 1995, White focused on a fossil that is one of the earliest members of the human family. He knew then that when he finished assembling the chalky fragments of this partial skeleton in the lab, fitting them together like pieces in a three-dimensional puzzle that is missing half of its parts, he would be the first person to glimpse an early human ancestor that was alive more than 4 million years ago. For years, he and other paleoanthropologists could only imagine what the earliest members of the human family looked like. Here, covered in silt, was the real thing. As he removed the silt from the fossil, millimeter by millimeter, he was uncovering the outline of a primate that was already offering surprises.

Just a few months earlier, in September 1994, White and his colleagues had introduced the first fossils they had found of this type of early human ancestor on the cover of *Nature*. This was a career achievement for a scientist, akin to getting on the cover of *Time* or *Newsweek*—usually only groundbreaking research or discoveries get such prominent play. In the same way that politicians display photos of themselves posing with prominent colleagues on their walls, White has framed photos of famous fossils he has known that were featured on the covers of *Nature, Science,* and other leading journals. The most famous of those is the fossil of a small jaw that appeared on the cover of *Nature* in 1994, showing a milk tooth of what they called an entirely new type of hominid—one that White and his colleagues initially named *Australopithecus ramidus,* a name that honored the local Afar people, *ramidus* in their language meaning "root," suggesting that it was humanity's root ancestor. At an age of 4.4 million years, this fossil was the first in twenty years to challenge Lucy for her status as the earliest human ancestor.

That claim to being the earliest ancestor, however, depended on seventeen fossils, most of which were teeth or jawbone. That meant *A. ramidus* had earned its place on the human line by the skin of its teeth, as the paleoanthropologist Bernard Wood of George Washington University, a frequent commentator in *Nature,* wrote in an accompanying article on the new fossils. Although Wood exulted that it looked to him like the long-sought missing link with the ancestor of chimpanzees, others said White had not proved beyond a doubt that *A. ramidus* was a hominid—he needed bones from the neck down to show how this creature moved through the prehistoric woods where it lived. To pass the traditional litmus test for being a bona fide hominid, it had to walk upright, rather than on its knuckles like a chimpanzee or gorilla.

So in November 1994, immediately after the rains stopped and the dry season started in the Middle Awash, White and his colleagues were back at work in the fossil beds near the village of Aramis, 150 miles northeast of Addis Ababa. There, less than two hundred feet from where they had made their initial discovery of the teeth of one individual of *A. ramidus* in 1992, the team members crawled across a shallow gully, with gloves on their hands and pads on their knees, positioned shoulder to shoulder so they would not miss a fragment in the rubble. As they moved up an embankment, the recent rains had eroded the surface, exposing a fragment of bone camouflaged by dust and rocks. A reserved, soft-spoken Ethiopian graduate student with a talent for finding fossils, Yohannes Haile-Selassie, was the first to spot two pieces of a bone from the palm of a hand. That hand bone was followed by pieces of a pelvis; leg, ankle, and foot bones; arm, wrist, and hand bones; a lower jaw with teeth—and a skull. None of the pieces were duplicates, which told them they had found the partial skeleton of a single individual. Before this discovery, only a few other partial skeletons of hominids that were older than 1 million years had been discovered, and the oldest were Lucy's remains.

The excitement was tempered, however, by the condition of the skeleton. The bone was so soft and crushed that White later described it as roadkill. It would take the Middle Awash team three years to excavate it from the soil and rock. After the team members managed to remove large blocks of sediment with embedded fossils, they would drive the blocks at the end of the season to Addis Ababa, where White would finish the excavation in the lab at the National Museum.

White had just returned from the field with parts of the ancient skeleton in January 1995 and was still in the early stages of removing the fossils from their blocks of hardened sediment when he got a visit from Meave Leakey. Leakey is a zoologist whose expertise is in fossils of monkeys. She had come to Addis Ababa with casts of fossils discovered a few months earlier on the shores of an ancient lake on the western side of Lake Turkana, a crescent of green water that cuts 180 miles from the border of Ethiopia into northern Kenya. Leakey was there to compare the fossils her team had found with those being analyzed by White and his colleagues to see if these extinct creatures showed any signs of being related to each other.

Meave had known White since he was a graduate student at the University of Michigan, and was undoubtedly eager to see the fossils. She is also the wife of famed fossil hunter Richard Leakey, and by 1995, when she made her trip to Addis Ababa, she was in the awkward position of knowing that Richard and White had not spoken to each other in years. The Leakeys met White on his first trip to Kenya, in 1974. They had invited him to work with the African team that found fossils at Koobi Fora, on the eastern shore of Lake Turkana; he stayed for three field seasons. At the time, Richard, Meave, and their famous team of fossil hunters, called the Hominid Gang, were known for their discoveries of a stunning series of hominids that lived near the lake almost 2 million years ago. In a remarkable dynastic succession, Richard had inherited the mantle of being the world's most famous fossil hunter from his father, Louis Leakey, who was the first to show that early humans evolved in eastern Africa.

An early photo of Meave and Richard in *National Geographic* in 1970—the year they were married—showed them both astride camels, setting off into the desert. Richard had organized the camel safari with an obvious eye for the photogenic. He sat atop a wooden saddle covered with sheepskin, with a red-and-black turban on his head and a broad smile on his tanned face. Meave was on her own camel beside him, lithe and fair. Never mind that the camels proved to be more trouble than they were worth, stubborn, slow, and a lure for lions and desert bandits. The image suggested nineteenth-century explorers—and may have prompted White's disparaging remark that fossils are not found according to *National Geographic* mythology. Richard and Meave were indeed tall, bronzed, and good-looking, as if they had been cast for the parts, and were having the time of their lives. This article was the first of a long series in *National Geographic, Life,* and other publications. They were a golden couple with a certain mystique. Richard even joked that in those early days, fossils practically fell out of the ground when they approached, as if their appeal was magnetic even to long-dead human ancestors. Their camp at Koobi Fora also drew a steady stream of young scientists, wealthy philanthropists, ambassadors, and even a Dutch prince and his entourage.

In truth, Meave did not bask in the spotlight. She was at heart a serious zoologist with a Ph.D. who was happier lying on the ground at eye level with a fossil than facing a camera. She has an extraordinary adventurous spirit and work ethic and enjoys the fossil talk and storytelling around the camp table at the end of a long day in the field. But she had not traveled from her native Wales to Africa to search for the missing link. Meave was intrigued instead by the evolution of animals in general. Though she recognized that the hominids were special and brought in funding for more expeditions, she was not infected with the hominid fever that swept through Koobi Fora, which became a training ground for a long line of researchers. Many were competitive young men like White, who stood out as particularly bright and aggressive even in this group. Unlike Meave and many of these young scientists, Richard was self-trained and had no academic degrees—and was proud of it. He had absorbed a lot from his parents and was a natural leader who knew how to run a camp and find fossils in the most hostile terrain. He was very much the magnetic center of this paleo-universe, surrounded by an inner orbit of loyal paleoanthropologists and anatomists working with him on the prize hominid fossils, even as he came and went to attend to museum business in Nairobi and to raise funds from international donors.

In this crowd, Meave was a low-key presence. She once joked during a public lecture that she got to work on the hominids only when Richard was distracted. She showed a slide of herself hovering behind Richard, trying to get a glimpse of a fossil he was examining. In fact, she was as serious as anyone about her research and had spent thousands of hours over the years finding, excavating, and examining fossils of hominids, as well as of other animals. Photos of those early expeditions often show her bent over fossils at a wooden table in camp, gluing tiny fragments of skulls or other shattered bones back together like pieces of broken porcelain. Not surprisingly, she and Richard talked fossils night after night at dinner at home. Meave would recall years later how much this would annoy their two daughters, who she also brought to camp, first as infants and then as young girls; as soon as they were old enough, they would help fetch water and work on the excavations. Yet though Meave was contributing to the team in significant ways and was respected for her perseverance and her knowledge of primate fossils, she—like many others—spent two decades in the shadow of Richard's brilliant blaze.

That changed rather suddenly on April 20, 1989, when it was announced on the radio in Kenya that President Daniel arap Moi had appointed Richard to take over management of Kenya's Wildlife Service, effective immediately. Richard had been planning an expedition to west Turkana that summer, but he immediately resigned his position at the National Museums of Kenya to take on the challenge of rescuing the national parks. Poachers had shot hundreds of elephants and had even killed game wardens and some tourists. Saving wildlife and restoring order to the parks was a cause close to Richard's heart, and it came at a time when he was ready for a new challenge.

Not only was he highly visible and successful, which made him the object of envy and emulation, but he also had tremendous power to recommend who got access to fossils and fossil sites in Kenya. His decisions—and autocratic approach—made him enemies. He had followed in his father's footsteps, inheriting leadership of the National Museums of Kenya, which they built over the years into a world-class research center. He had established rules and de facto practices that would influence how many expeditions would work in Kenya for years to come—but some people chafed under his leadership. One of his most vocal critics, in fact, was White. An exchange of letters between White and Richard only intensified, rather than resolved, the tension—and actually ended in threats of legal action. Richard complained years later that the criticism from White and others plagued him even when he was in the hospital in 1979 and early 1980, fighting for his life after a kidney transplant. Almost a decade later, at the age of forty-five, he was weary of the highly personal politics of paleoanthropology and ready to step away from the field for a while.

Meave was prepared to take charge. She had already become head of paleontology at the National Museums and was making plans to head the field expedition to west Turkana that summer. She and Richard and their colleagues had spent two decades focusing on fossil beds at Lake Turkana that were 1 million to 3.5 million years old. It was time, she decided, to look for fossils in older sites.

By the time she traveled to Addis Ababa to see White in January 1995, she was fifty-three years old and had spent several months every year for five years doing backbreaking work in the scorching sandstone exposures of the west Turkana desert, searching for fossils that were more than 4 million years old. Her straight blond hair was silver-gray now, and her face was lined from years in the dry desert without sunblock or cover. She had weathered difficulties that would have discouraged many others. Yet over the years she had acquired an air of quiet resolve and strength, like steel tempered by fire. Her first expeditions to west Turkana had proved disappointing—in four years of fieldwork at a place called Lothagam, her team had found fossils of every imaginable extinct animal, but only six teeth of hominids. During the first week of her last season there, Richard had nearly been killed in a plane crash. His road to recovery had been a difficult one, involving the amputation of both of his legs in England and many months of physical therapy to learn to walk again. Finally, in May 1994, Meave could return to the field. She moved the team to Kanapoi, an ancient lakebed in the undulating badlands of southwest Turkana. It was time for a fresh start.

There, just two weeks after setting up camp along a dry riverbed, her team scored a major discovery. She would describe it in *National Geographic* a year later: one of the members of the Hominid Gang, Wambua Mangao, called out excitedly to her; she followed him to a slope glazed with volcanic rubble. There, she could see bits of bluish tooth enamel embedded in a rock. When she turned the rock over, she realized it held half of the upper jaw of an animal the size of a chimpanzee. It was a hominid, but what type of early human? Could it be the earliest known hominid? She knew from the geology that it was about 4 million years old. But she was unaware that a member of White's team was in London that very month, hand-delivering two manuscripts to an editor from *Nature,* describing their new fossils from Ethiopia that were 4.4 million years old.

Over the next few weeks, the renowned Hominid Gang fossil hunter Kamoya Kimeu found two parts of a shinbone that resembled those of Lucy's species, showing that this chimp-sized creature had walked upright. Then came more teeth and, on the last weekend of the field season, a complete lower jaw and the ear region of a skull. These body parts from several individuals showed a mixture of primitive traits, such as those found in a chimpanzee, and more derived characters—new traits that appeared later in early humans but not in the African apes. The new fossils also showed some unique features of their own. The mix suggested that Meave and the Hominid Gang had assembled fossils of a new type of early human—an upright-walking hominid that was more primitive than Lucy.

By the time she returned to Nairobi, Meave was delighted with the finds they would soon report. She was convinced that her team had collected the most complete known specimens of a hominid this old. Unaware of the discoveries made by White and the Middle Awash Research Group in

Ethiopia, Meave thought she might even have in her hands the remains of the oldest known hominid, and a possible ancestor for Lucy's species.

Then she learned from the Leakey biographer Virginia Morell that White was about to announce a new hominid species from Ethiopia that was even older than her fossils from Kanapoi. Before she even had announced her team's discovery of the earliest known hominid, her fossils already had been eclipsed by those discovered by Tim White and the Middle Awash group. When the reports appeared in *Nature* on September 22, 1994, she studied the photos and scoured the description of the fossils. Could the new species they were announcing, *A. ramidus,* be related to the fossils she had found? What was the evidence that *A. ramidus* was even a hominid?

Meave wrote to White about her team's discoveries at Kanapoi, and he invited her to Addis Ababa so they could compare fossils. It was in January 1995 that the two met in a single-story concrete-block building—a lab built with American National Science Foundation research dollars that was torn down in 2004 to make room for a new, five-story laboratory. They met in the spartan hominid room, where White brought out fossils of Lucy's kind, *A. afarensis,* to see how they compared with the casts of fossils that Meave had brought. Then, one at a time, White showed Meave original fossils of *Ardipithecus ramidus* from the safe, comparing teeth from *Ardipithecus* and Meave's fossils.

What Meave and White saw would lead them to conclude that her fossils were a new kind of early human—different from the fossils White and his colleagues had found. They were pleased that the Kanapoi fossils appeared to be intermediate between *A. ramidus* and Lucy's species, both in time and the way they looked, perhaps showing one lineage evolving over time. White told Meave that she should focus on finding baby teeth—in particular, lower molars that could be compared directly with the particularly informative milk molar of *A. ramidus.* (She would find such molars, and it would confirm their sense that this was one lineage.) The meeting was productive and friendly, Meave remembered years later. White was hospitable; he even paid her hotel bill.

Later that year Meave and paleontologist Alan Walker and their colleagues published their description of the fossils at Kanapoi and those from another bone bed on the eastern shore of Lake Turkana called Allia Bay, where they had found more fossils of the same type. They named them *Australopithecus anamensis,* drawing on the Turkana word *anam* for "lake," which was fitting since all of the fossils came from the floodplain of a river that flowed into the ancient Lonyumun Lake, whose waters extended far beyond the shores of modern Lake Turkana. They would propose this "ape of the lake" as the direct ancestor of Lucy's species. The shinbone provided the oldest direct evidence for upright walking, making *A. anamensis* the earliest indisputable hominid. The editors at *Nature* commented that these were exciting times for paleoanthropology. Meave and Walker had heard that White's team had excavated even more fossils of *A. ramidus,* and they were wondering if the bones from the neck down proved its identity as an upright-walking hominid. "We're on tenterhooks waiting to see what Tim has," Walker told the *New York Times.*

Ancient

Footsteps

I hear the ancient footsteps like the motion of the sea
Sometimes I turn, there's someone there, other
times it's only me.
I am hanging in the balance of the reality of man
Like every sparrow falling, like every grain of sand.

BOB DYLAN

CHAPTER ONE

AFRICAN TRAILBLAZERS

Most scientific problems are far better understood by studying their history than their logic.

ERNST MAYR, evolutionary biologist

I was told as a young student not to waste my time searching for Early Man in Africa, since "everyone knew he had started in Asia."

LOUIS LEAKEY, 1966

It was an October morning in 2003. Meave Leakey was driving from Nairobi north along the eastern wall of the Rift Valley in central Kenya, expertly weaving around potholes in the tarmac and dodging oncoming buses that played chicken with smaller vehicles to scare them out of their way. Trucks belched black smoke that stung her eyes, cyclists hitched rides up hills holding on to the backs of buses, and jam-packed public shuttles called *matatus* spent almost as much time passing each other as staying on their side of the two-lane road. As Meave negotiated this nerve-racking traffic on the Uplands Road between Nairobi and Nakuru, she calmly recounted the story of how the search for human ancestors began in eastern Africa. "Until the middle of 1959, only a few people seriously believed eastern Africa was a sensible place to look for the earliest human ancestors," she said.

This history is personal for her, because it is the saga of her husband's parents, Louis and Mary Leakey. This formidable pair was among the first to stake their careers on Africa as the birthplace of mankind. For three decades, their work in eastern Africa was an almost solitary pursuit. Even those researchers who found fossils of early ape-men in South Africa during that time had trouble convincing their European colleagues that these primitive fossils were ancestors of humans. Then, in 1959, Mary found a fossil in Olduvai Gorge, Tanzania, that would finally give the Leakeys the hard evidence they had long sought that early humans did indeed evolve in eastern Africa. Louis named the cranium, or partial skull, *Zinjanthropus boisei*—*Zinj* from an Arabic word for eastern Africa, *anthropus* from the Greek word for man, and *boisei* from Charles Boise, a London businessman who was their benefactor. Translated, the name is an assertion: "Man from Eastern Africa." And once the Man from Eastern Africa made his appearance, the push was on to find more extinct men and women. Soon, teams of French and American researchers headed to eastern Africa, like forty-niners to California during the gold rush. The fossils they found in the Great Rift Valley in the 1960s and 1970s soon made it known as the cradle of humanity.

But Louis's search for the missing link in eastern Africa had started more than thirty years earlier, right in the gullies and rock shelters alongside the Uplands Road where Meave was driving nearly eighty years later, high above the Great Rift Valley. In 1926, the first year that Louis worked in the area, the Uplands Road did not exist, and the trip from Nairobi in Louis's Model T Ford took a half day over muddy tracks. The air was so clear that Louis could see miles across the Great Rift Valley from his camp, down a slope covered with acacia trees and scrub brush to Lake Elmenteita, a shallow alkaline lake rimmed with the pink froth of flamingos. Bush babies, leopards, aardvarks, and ibises lived in the acacia woodlands near the shore, and a herd of hippos wallowed in the lake. Beyond the lake, the jagged calderas of several extinct volcanoes lined up to form the silhouette of a human figure that the local Masai tribesmen called Elngiragata Olmorani, for Sleeping Warrior. A few British settlers were staking out the Masai's traditional grazing grounds for homesteads for cattle ranches, but otherwise the area was still remote and primeval.

Today, Elmenteita is only an hour's drive beyond the shanty sprawl surrounding Nairobi, and much of the land around the lake is fenced in by private owners. The hippos are gone and the bush babies and a few remaining leopards have retreated to a wildlife sanctuary. But the view of the rift valley far below is still stunning, and Meave named the volcanoes visible in the distance as she searched for a familiar turnoff. Spotting it, she jostled down a dirt road, past a quarry where workers mined a crumbly white rock called diatomite, and pulled into a grassy driveway. The sign said: KARIANDUSI MUSEUM, NATIONAL MUSEUMS OF KENYA. It did not look like much: a guard's hut and a whitewashed, single-room museum with some casts of skulls and an exhibit on the formation of the Great Rift Valley. It was clearly off the tourists' safari circuit.

A curator eventually appeared, delighted to find someone who wanted to tour the site on a Monday in October. He was even more surprised to find out that the tall woman with straight, silver-gray hair and hazel eyes spoke Swahili and was a member of the Leakey family—a name that is well-known in Kenya. At sixty-one, Meave had been here many times before and knew the history of Kariandusi by heart. She is long-legged and fit after a lifetime of hard work scrambling over rugged terrain for fossils, and she did not need a guide to lead her into the gulch. She let the curator show her the way to a series of steplike pits anyway, partly because she was curious to learn what he knew. He took her to the first pit, which was covered with a corrugated metal roof. Meave leaned over the rail and pointed to the dirt floor encrusted with hundreds of stone tools, most made of glassy black obsidian, the rock that comes from volcanic lava. There were tear-shaped hand axes, two-sided flakes, and even triplets of round stones that look like black billiard balls. "This is where it all began," said Meave. She was referring to Louis's search for early man in eastern Africa. These shiny black tools at Kariandusi were among the first hard evidence of a sophisticated ancient Stone Age culture in eastern Africa. They were made by people who left them on the shores of the lake almost 500,000 years ago, perhaps when they came to hunt wild animals that were quenching their thirst at dawn or dusk. She climbed down wooden stairs into a deep gully where an ibis was roosting in a tree. Meave remembered a photo from *National Geographic* that showed Louis bending over a cliff there, pointing to stone tools embedded in the wall. This was precisely the spot where Louis and his team found their first ancient hand axes in eastern Africa in 1929.

In 1871, less than sixty years earlier, Charles Darwin had proposed in *The Descent of Man and Selection in Relation to Sex,* that the earliest ancestors of humans probably lived on the African continent. But that prediction was based on absolutely no evidence from fossils. In fact, at the time only one fossil of another type of human being was known, and that was of a Neandertal that had lived in the Neander valley of Germany sometime in the past 70,000 years. Darwin chose Africa because humans' closest cousins in the animal kingdom—chimpanzees and gorillas—lived in Africa; therefore, he wrote, "it is more probable that our early progenitors lived on the African continent than elsewhere." But Darwin admitted that it was "useless to speculate on this subject," since an extinct European ape nearly as large as humans could also have given rise to humans.

That didn't stop Darwin's colleagues from conjecture. His friend and champion Thomas Henry Huxley (also known as Darwin's "bulldog") agreed that humans should be put in the same family as chimpanzees and gorillas, and enthusiastically promoted that view in debates and in his 1863 book *Evidence as to Man's Place in Nature.* (Darwin himself avoided dealing directly with the issue until 1871, when he published *The Descent of Man.)* But a contemporary and admirer of Darwin's, the prominent German biologist Ernst Haeckel, believed that the Asian apes (orangutans and gibbons) were closer relatives of humans than the African apes were. Haeckel proposed this link in his sketches of the human family tree in 1868, drawing a direct line between Asian apes and a new species of fossil human that he proposed and explicitly called the missing link. In his writings and lectures, Haeckel fleshed out this missing link as a hairy, primitive creature half ape, half man, named *Pithecan-thropus alalus.* (Literally, "ape-man without speech," from the Greek *pithec,* "ape," *anthropus,* "man," and *alalus,* "without speech.") It walked semierect, had protruding teeth, and was speechless. But there wasn't a bit of hard evidence to support this vision of an ape-man. Haeckel's missing link was purely theoretical.

One person who heard of Haeckel's ideas on human evolution was a young Dutch medical student, Eugène Dubois, who became the first of a long line of young men obsessed with finding this missing link—and winning honor and fortune. In 1887, when Dubois couldn't get the Dutch government to finance an expedition to the tropics to search for fossils, he quit his job as an anatomist at the University of Amsterdam and joined the Royal Dutch East Indies Army as a military doctor so he could be posted to the Dutch East Indies, now the Indonesian archipelago. Ancient fossils of mammals that had been alive during the earliest stages of the Age of Man (the Pleistocene epoch) had been found there. He thought it most likely that fossils of extinct ancestors of similar age would be preserved there as well. According to his biographer, the anthropologist Pat Shipman, he also reasoned that if apes lived in the tropics today, extinct apes and early ape-men would also have been more likely to live in the tropics.

It was an incredible long shot, but he sailed for the Dutch East Indies at the age of twenty-nine with his young wife and their baby. Dubois was the first of many fossil hunters to risk his life in search of an elusive missing link. He battled malarial fever without modern medicines; his team hemorrhaged workers, who ran away, became ill, or stole fossils to sell as "dragon" bones to traders from China; and they faced bad roads through the overgrown jungles of Java, mosquitoes, hellish heat, and torrential rains. Amazingly, Dubois and his family survived. More incredibly, he found what he was looking for. In August 1891, his crew discovered the molar of a hominid eroding out of the banks of the Solo River near the village of Trinil on the island of Java. Two months later, his crew found a skullcap that was larger than that of a chimpanzee's but smaller than that of a human's. Later, they found a thighbone. Dubois recognized the skullcap as belonging to a species that must have had a brain intermediate in size and development between humans and apes. But the thighbone belonged to a creature that walked upright—even before its brain had expanded.

He pronounced it *Pithecanthropus erectus* (or "erect ape-man"). It was an amazing feat. He had never searched for fossils but had nonetheless traveled halfway around the world to an island archipelago where he'd *reasoned* that such fossils should be found. Today, Dubois's Java man is still recognized as a major discovery—the first fossil found of an early hominid and the first specimen of *Homo erectus* (as it was later renamed), a key human ancestor that arose about 1.8 million years ago, probably in Africa, before migrating to Asia, where it persisted until sometime in the past 250,000 years. This species of human and its descendants may even have lived until as recently as 13,000 years ago in the form of the so-called Hobbit, the dwarf species of human whose remains were found in 2004 on the Indonesian island of Flores.

Convincing his colleagues that he had found the missing link would prove more difficult than finding the fossils themselves. When Dubois announced his discovery of Java man in 1893, he expected honor and scientific recognition. Instead, his monograph on this "man-ape" was met with skepticism and snide comments, some dismissing the fossil as a giant gibbon or an individual whose features had been distorted by disease or a wound. Word reached him in Java in 1894 that his European colleagues questioned many aspects of his monograph on the fossil—from his claim that all the fossils came from the same individual to the way his crew had mapped the fossil site.

Dubois traveled to Europe in 1895 to defend his discovery, winning a few converts as he lectured and displayed the bones themselves. Haeckel, who had inspired him, was one who embraced Java man as a human ancestor. But the theory of evolution was still new and was not universally accepted among scholars. Although Dubois was well educated and a meticulous scientist, perhaps the real problem was that an ancestor that looked so much like an ape was more than the scientific establishment of the late nineteenth century could accept. His biographer Shipman concluded, "In truth, the problem lay more in the prevailing beliefs among his colleagues than in Dubois' shortcomings."

As he battled his colleagues well into the twentieth century, Dubois's own shortcomings also became apparent—he grew secretive and territorial about his fossils, particularly after he gave a cast to a German anatomist who then toured the world with it, giving lectures and publishing a detailed description about Java man before Dubois had finished his own analysis of the skull he had found. After that, he withdrew from his colleagues and even rigged a mirror above his door at home so he could see who was there when his maid answered, turning away prominent scientists who'd traveled from as far as America to see the fossils that he stored in his basement.

History would prove Dubois right about Java man, but he died an angry man, unrecognized and estranged from his wife and friends—all alienated by his increasing irascibility. He was, perhaps, the first fossil hunter to become a victim of his own success in finding a human ancestor, as if the fossil came with a mummy's curse.

It was a bitter omen of the kind of controversy that would swirl around almost every new fossil vying to be a human ancestor. Even experienced researchers often react with more emotion to the discovery of human ancestors than they do to fossils of any other animal, including dinosaurs. New fossils almost always shatter preconceived notions of what our ancestors should look like, revealing our origins as ordinary apes rather than as exalted beings marked from the beginning with a big brain or some other sign of special destiny. Darwin recognized this reflexive denial of our savage past in *The Descent of Man* when he warned, "We must, however, acknowledge, as it seems to me, that man with all his noble qualities, with sympathy which feels for the most debased, with benevolence which extends not only to other men but to the humblest living creature, with his god-like intellect which has penetrated into the movements and constitution of the solar system—with all these exalted powers—Man still bears in his bodily frame the indelible stamp of his lowly origin."

Preconceptions of what the missing link should look like, in fact, colored the scientific community's remarkable acceptance of the next fossil proposed as an early hominid, almost twenty years later. This was the Piltdown hoax, a sorry tale that began when a young English soldier and amateur anti-quarian, Charles Dawson, "discovered" fragments of a skull, a lower jaw, and some teeth in a gravel pit at Piltdown in Sussex, England, over the years be-tween 1908 and 1915. Dawson's fossils were convincing to several leading sci-entists of the day, including paleontologist Sir Arthur Smith Woodward, who was keeper of geology at the British Museum, and Sir Arthur Keith, the most eminent British anatomist of the day. Keith had spent a year studying the skulls of two hundred primates and come to the conclusion that the essence of being human was to have a big brain relative to the size of the body.

With Piltdown man, Keith got exactly what he was looking for. When put together, the big brain and primitive apelike jaw of Piltdown precisely fit Keith's image of a "missing link" between humans and apes. Keith embraced it as a direct human ancestor, and defended it for decades. There were some skeptics, chiefly Americans who argued that the jaw did not belong with the skull.

It turns out that Piltdown man *had* been made to order. Scientists from Oxford University and the British Museum formally declared it a fake in 1953 after chemical tests showed that the jawbone belonged to an orangutan and that the skullcap was human and older than the jawbone. The forgers had filed the molars to look more human, and they had stained the bones and stone tools to match before planting the "fossils" in the gravel pit. They were obvi-ously familiar with Keith's notion of a human ancestor. Dawson is the front-runner in a line of suspects, but he died in 1916, perhaps before he planned to admit to the hoax.

Unfortunately, Piltdown man continued to cause damage long after Dawson died. For forty years, in fact, fallout from Piltdown would delay the recognition of the authentic fossils, including Dubois's Java man. Java man and fossils found later in South Africa did not fit the image as well as Piltdown for the one missing link between modern humans and apes, so they were rel-egated to side branches on the family tree—shown as separate species, types or "races" of humanlike creatures that had gone extinct. Worse, some were even thought to have given rise to humans still living in the Stone Age, but not Europeans. Confusion persisted for many years. One popular book, called

Meet Your Ancestors, by Roy Chapman Andrews, a retired director of the American Museum of Natural History, vividly illustrates a popular view of the human family that still persisted in 1945, when the book was published. On the book's endpapers, a diagram labeled "The Family Tree of Man" shows Java man giving rise to living Australian aborigines. Another Chinese fossil that was considered a close cousin of Java man begets living Mongoloids. Both of those lineages start with a ground ape. But a separate lineage that originated with a forest ape leads to the European cave painters known as the Cro-Magnon, whom the author describes as "wise men" and the ancestors of European *Homo sapiens.* Andrews allows for some inbreeding but describes aborigines as "not much advanced beyond the stage of Neanderthal Man." The message was clear: Europeans were a superior race with their own largely separate ancestry from forest apes, rather than from lowly ground apes.

⁘

The most significant fossil eclipsed by the long shadow cast by Piltdown was the first early hominid skull from Africa. In 1924, an Australian anatomist named Raymond Dart had taken the post of chair of anatomy at the University of Witswatersrand in Johannesburg, South Africa, after completing his medical studies in London. When he got to Johannesburg, he was dismayed to find no reference collection of skeletons to study. So he announced a competition to see which students could bring in the most interesting bones. He described the outcome years later in his autobiography, *Adventures with the Missing Link:* One student, the only woman in the class, told him she had seen the skull of a fossil baboon on the mantel at a friend's house. Dart told her it was probably not a baboon, since he knew of only two fossil primates that had been found at that time in sub-Saharan Africa. The student borrowed the skull, which was indeed a fossil baboon. It belonged to her friend, a director of the Northern Lime Company at Taung, where fossils turned up occasionally in the limestone rock. Dart asked the manager to send him any other fossils they might find.

Later that year, in the summer of 1924, two wooden crates were left on

Dart's stoop, just as he was struggling to put on a stiff-winged collar for a friend's wedding at his house, in which he was to serve as best man. Dart tore off the collar and pried open the crates with a crowbar, against the admonishments of his wife, who warned that the wedding guests would soon be arriving. In the first crate, there was little of interest. But in the second crate, right on top was the fossilized cast of a brain in limestone—an endocast, which is the imprint of the brain on the inside of the skull, much like a death mask is the plaster impression of a face. This one was complete with the pattern of blood vessels. Having made an intensive study of endocranial casts during his medical studies in London, Dart was thrilled. Ransacking the box for the face that matched the brain cast, he quickly found a large piece of rock and fossilized bone that fit perfectly. He could see that the brain came from a creature with a forebrain that was larger than a chimpanzee's, although it was not big enough to belong to a primitive human. He was standing in the shade admiring it when the groom found him and broke his reverie, telling him he had to finish dressing because the guests were arriving. But, Dart would write in his autobiography, he could scarcely wait until the wedding was over so he could reexamine his "treasures," which he had stored in his wardrobe.

It took him months to clean the fossil, like a sculptor chiseling away stone from a form he sees in marble. Eventually, on December 23, he split the rock with a hammer and his wife's steel knitting needles, and the face emerged. What Dart saw was a baby's face, with a full set of milk teeth and its first permanent molars just in the process of erupting. "I doubt that there could have been any parent prouder of his offspring than I was of my Taungs baby on that Christmas of 1924," Dart wrote. The baby was neither ape nor human. It had small canines like a four-year-old human, and the position of the head right on top of the spine indicated that it had walked upright. A neuroanatomist who had not set out to find the missing link, Dart soon realized that he had found a probable early member of the human family. By early 1925, Dart was ready to baptize his baby: he published a paper in the prestigious journal *Nature*, in which he named it *Australopithecus africanus*, or "southern ape from Africa." It soon became known as the Taung baby. A few immediately recognized its importance as a hominid that was even more primitive than Dubois's Java man.

Others, perhaps blinded by the anatomy of Piltdown, dismissed the Taung baby as a young ape, rather than a hominid. It had exactly the opposite gestalt of Piltdown: where Piltdown had a large brain and primitive jaw, Taung had a relatively small brain (although the forebrain was noticeably large) and modern teeth. Both could not have been direct ancestors of modern humans. Keith suggested that Dart had mistakenly identified the Taung baby as a hominid because its young age made it look more modern. In time, he wrote, its face and teeth undoubtedly would have looked more apelike as it developed into an adult ape.

Like Dubois, Dart also traveled to England to rescue the reputation of his fossil hominid. He had anticipated skepticism, but when he got there in 1931, he was not prepared for his delicate skull to be upstaged by a new fossil. This one was from China—Peking man, a skull discovered in 1929 by a young Canadian physician, Davidson Black. The skull resembled Dubois's Java man, but Black gave it a different name, *Sinanthropus pekinensis,* that would persist for years before paleoanthropologists recognized that it and Java man were the same species—the renamed *Homo erectus.* Black was much more effective than Dart or Dubois in spreading the news—sending tactful advance warning to leading scholars he had trained with in England, then touring Europe and the United States immediately with casts and a slide show, which helped build the case that Peking man was a serious contender for the missing link. He claimed that his fossils were intermediate between Dubois's Java man and Neandertals, putting Peking man a step closer to being human.

By the time Dart appeared in England, his presentation on the Taung baby was anticlimactic, and the scientists he visited were much more interested in talking about the Chinese fossils they had seen and an Asian origin for humans. The Taung baby was a nice fossil, but it was an also-ran, not on the one line of descent to humans. Keith's criticism of the Taung fossil even extended to Dart—he later wrote that he was "rather frightened" of Dart because of "his flightiness, his scorn for accepted opinion, the unorthodoxy of his outlook." Personal politics again colored the reception of a bona fide fossil hominid.

When Dart left London, he was deeply discouraged. He did not fight back. Instead, he put aside his monograph on the fossil and even gave up work on other fossils for many years. He later would explain that he had no burning zeal for fossils and had not come to Africa to search for the missing link. He had other professional interests in neurology and was by then eager to get out of the field of human origins.

⋙

Dart's experience should have discouraged any young researcher who hoped to find a missing link in Africa. But Louis Leakey, who was a student at St. John's College in Cambridge, England, liked to challenge conventional wisdom and had unshakable confidence in his own instincts. He had read Darwin's *Descent of Man*, knew about the Taung baby, and had heard about the discovery of a fossil skeleton in 1913 at Oldoway in the British colony Tanganyika—now Olduvai in the nation of Tanzania. These African fossils reinforced his own feelings that Africa was the birthplace of humankind. "I became excited with the idea that everyone was looking in the wrong place," he wrote in his first autobiography, *White African*. And he began to make plans to lead his own small expedition to eastern Africa in the summer of 1926, where he intended to excavate a cave that he and his sisters had discovered when he was a young child growing up in Kabete, in the lush mountains north of Nairobi, where he was born.

Louis was the son of British parents who ran an Anglican mission in Kabete, and he had acquired a missionary's zeal and single-mindedness of purpose in the face of obstacles that would have daunted most men. As a boy, playing with his friends from the Kikuyu tribe, he had found obsidian stone tools in the countryside around Kabete. He had taken some of those tools to a zoologist at Nairobi's Coryndon Museum who was a friend of the family's and who had taught Louis how to classify birds and label museum specimens. The zoologist confirmed that the tools Louis had found were indeed arrowheads. Louis decided then that he wanted to be both a missionary, like his father, and a scientist, like the zoologist. He would one day find out about the Stone Age men who had made the tools he found. At the age of thirteen, he had already set his course.

But when he was sent to boarding school in England at age sixteen, he faced challenges that threatened to derail his plans. Louis later wrote that when he went to school, he was shy and unsophisticated and fit badly into the life of the place. His friends in Kenya had been Kikuyu boys who considered him a blood brother, because he had taken part in their secret initiation ceremonies. Though he had studied Latin and mathematics, and had spoken French at the dinner table in Kenya, Louis related, "In language and mental outlook I was more Kikuyu than English." He did not know Greek; worse, he had never played cricket and did not know how to swim, which brought him ridicule from much younger boys and hazing from older boys. In Kenya, he had been used to a great deal of freedom, building, near his parents' home, his own three-room house, where he lived as a teenager. He earned money by trapping animals and helped his father teach school in Kabete. In England, he chafed at the rules for bedtime and passes for visiting the nearby town, which made him feel like "a child of 10 when I felt like a man of 20." He even had to get permission to stay up late to catch up on his studies. Despite this effort, his headmaster discouraged his ambition to win a scholarship to attend Cambridge University, his father's alma mater.

Perhaps the experience of being different from the other boys contributed to the making of a maverick, because by the time he did get a small scholarship, to St. John's College at Cambridge, he already was charting his own course—convincing administrators to accept Kikuyu as one of two modern-language requirements. He also managed to take time off to search for dinosaur fossils in Tanzania while recovering from a head injury suffered while playing rugby. He got in trouble with the authorities for playing African signal drums on the roof of his rooms in St. John's College, but he earned top marks in anthropology and archaeology and got a small grant to spend a Christmas vacation studying collections of ancient artifacts in museums in Europe. While there, he met the prominent German paleontologist Hans Reck, who had found the skeleton in Olduvai. Though the skeleton would eventually prove to be the remains of a modern human that had intruded into older sediments, Louis half seriously made plans that one day they would visit Olduvai together.

As he studied the artifacts from ancient cultures in Europe, Louis became even more certain that there must be evidence of humans in Africa at least as old as those in Europe. At the time, archaeologists thought that the oldest human culture was one found in Europe—a stone-hand-axe tradition they called the Chellean that had been made by the earliest ape-men. Louis thought he could find similarly ancient stone tools and, perhaps, their toolmakers in Africa: "I was born in Eastern Africa and I've already found traces of early man there. Furthermore, I'm convinced that Africa, not Asia, is the cradle of mankind." One of his professors tried to dissuade him, uttering the now legendary admonishment that he should not waste his time looking for early man in Africa. He should go instead to Asia, home of Java man and Peking man.

Louis ignored the advice, since he was intent on finding out if Darwin was right about Africa being the birthplace of humankind. In 1926, he was awarded a research fellowship by St. John's to return to Kenya for a year to lead the First Eastern African Archaeological Expedition (consisting of himself and a fellow Cambridge graduate). After finding little of importance in caves he had discovered as a child in Kabete, he set up camp in an abandoned pigsty on a farm on the slope above Lake Elmenteita where he had heard about a prehistoric burial mound. Eventually he moved to an abandoned farmhouse where he was joined by his first wife, Frida, and a half-dozen other researchers whose excavations uncovered skeletons of prehistoric humans and tools.

It took them three years, working in rock shelters and on exposed hillsides along the ridges above Lake Elmenteita and the Nakuru-Naivasha basin before they found the evidence Leakey was seeking of a truly ancient culture. Finally, in May 1929, just three weeks before the end of the 1928–29 field season, two members of Louis's team surveyed the gullies at Kariandusi and discovered a cache of tear-shaped hand axes. Louis was delighted: these tools were exactly what he had sought—relics of a culture that appeared to be almost as ancient as the Chellean culture of Europe. He knew they were ancient because they were found along the shorelines of a giant prehistoric lake. He thought the tools dated to the first ice age, which at the time was thought to be 40,000 to 50,000 years ago. Never one for understatement, Leakey wrote in *White African:* "The discovery was, therefore, of the very greatest importance."

Today, the hand axes at Kariandusi are thought to be 500,000 years old, rather than 50,000. But his estimate fit with the best geologic dates known at the time, which were estimated using relative dating, which determines the age of a layer of rock based on the fossils it contains and the rocks and fossils found above and below it. For many years, there was no clear idea of the absolute age of each rock layer or of the earth itself. Therefore, in 1926, the age of the earth was thought to be only 65 million years (now it is dated to be 4.5 billion years), and the age of mankind (the Pleistocene) was estimated at 500,000 years (today the Pleistocene is dated precisely to just under 2 million years). The tools were indeed old and one of the first concentrations found in eastern Africa of the Acheulean tradition, a tool kit that was the Swiss Army knife for *Homo erectus,* the species of Java man and Peking man. Early modern humans, now known as archaic *Homo sapiens,* also used the same type of tool kit, until someone finally invented a more sophisticated stone technology, which caught on about 250,000 years ago.

Louis's triumph was short-lived. His stone tools should have been the beginning of a brilliant academic career. They were not. Louis did go to Olduvai with Reck in 1931, and he also worked at two sites in western Kenya, Kanjera and Kanam, where he found a jawbone in 1932. His troubles began when he boldly claimed that the jawbone, which he named *Homo kanamensis,* was "not only the oldest human fragment from Africa, but the most ancient fragment of true *Homo* yet to be discovered anywhere in the world." At first, his colleagues congratulated him, assuming Louis's date of 500,000 years was correct for the jawbone. Louis, not yet thirty, got his first taste of fame.

He returned to Kanam with the eminent British geologist Percy

Boswell to prove the ancient age of his jawbone. Instead, he found to his horror that the iron pegs he had left in concrete to mark the spot had been removed, perhaps by the local Luo people for fishing spears. And the photos that Louis had taken of the site had not turned out, so it was impossible to confirm the precise spot where the jawbone was found and thereby date the sediments to nail down the jawbone's age.

When Boswell returned to England and reported to the Royal Society that Louis did not know exactly where he had found his fossil, Louis's reputation suffered. Later, when the fossils were shown to be much younger and Boswell published his doubts in a letter to the journal *Nature,* Louis was labeled as an enthusiast who worked too fast to be trusted as a careful scientist. It also hurt him that he had scandalized Cambridge colleagues by leaving his wife, Frida, in 1934, when she was pregnant with their second child. He had fallen in love with Mary Nicol, a twenty-year-old amateur archaeologist who was illustrating a book Louis was writing on the evolution of man and his culture. The two were living together in a village near Cambridge.

Louis married Mary just as his professional prospects were dimming. His expectations of a fellowship at Cambridge faded, and his ability to obtain grants to search for fossils in Africa deteriorated. The combination of Percy's report on his unreliable science at Kanam and his illicit affair with Mary led to his total ostracism from British academic circles. Louis pressed on anyway, taking Mary to Olduvai in Tanzania in 1935, where they would work off and on for the rest of their lives. There were only tiny grants, and Louis earned just enough money to support his growing family by writing books about his life, human prehistory, and the Kikuyu people. But he and Mary continued to explore and excavate archaeological and anthropological sites in Kenya. They found stone tools, teeth, and even the jaw of an extinct ape that had lived millions of years ago on the shores of Lake Victoria in Kenya.

It wasn't until more than a decade later, in 1948, that they made a discovery of international importance—the first skull of a fossil ape ever found. It was the partial skull, face, and jaw of an 18-million-year-old extinct ape called *Proconsul* from Rusinga Island in Lake Victoria, Kenya. Mary, who had found it, flew with it on her lap to England so that the prominent Oxford University anthropologist Wilfrid Le Gros Clark could examine it. News of the discovery traveled ahead of her to London, where newspaper headlines proclaimed, THE LEAKEYS FIND IMPORTANT FOSSIL-MAN ANCESTOR. The headlines were wrong. Although Louis thought *Proconsul* a possible ancestor for later apes and man, he knew it was far too primitive to be a human ancestor. Today it is seen as ancestral to both the lesser and the great apes that came later.

Le Gros Clark was an important friend of the Leakeys. He played a critical role in helping to convince a growing number of prehistorians that Africa was the birthplace of humankind. A year earlier, at the First Pan-African Congress of Prehistory in Nairobi, organized by Louis, Le Gros Clark had openly stated that the Taung skull and other new fossils of its species were indeed near human and belonged in the human family tree. He had traveled to South Africa before the congress and had met with Dart and Robert Broom. Broom, a Scots-born paleontologist who had been trained as a medical doctor, was curator of vertebrate fossils at the Transvaal Museum in South Africa, where he would carry on the paleontological work after Dart became discouraged in the 1930s and 1940s. He and his assistant, John T. Robinson, also found a diverse group of hominids, representing at least three different species, in caves in South Africa. Those fossils included adult specimens of the same species as the Taung baby.

When Le Gros Clark finally saw these fossils from South Africa, he recognized them as hominids. Le Gros, as he was called, gave them his official blessing, and his lead was quickly followed all over the world. The forged Piltdown fossils had recently been given a long-overdue deathblow in 1953 when chemical tests were finally done at the British Museum; the fossils were declared fakes. As Piltdown exited, the stage was cleared for the South African fossils and the entry of the next major player in the prehistory of Africa, the now-famous skull of *Zinjanthropus boisei* that changed Mary and Louis Leakey's lives for good.

<hr>

Mary Leakey discovered Zinj, as it was called, on the morning of July 17, 1959. Louis was in camp with a fever, and she set out by herself on a walk with her two Dalmatians to explore the fossil beds at Olduvai Gorge, where Louis had worked off and on since 1931. A scrap of bone poking out of the dirt caught her eye. She brushed away a little of the sediment and saw two teeth, black-brown and set in the curve of a jaw. She knew right away it was a hominid skull, and rushed back to camp to break the news to Louis. The Leakey biographer Virginia Morell described how Mary told Louis over and over: "I've got him! I've got him! I've got him!"

Louis, who was groggy with fever, was puzzled: "Got what?"

"Him, the man! Our man. The one we've been looking for," she replied.

Louis joked later that he quickly recovered from his fever as soon as he saw the fossil. He was disappointed to note that the skull had teeth just like those of the South African australopithecines, which he had considered an evolutionary side branch that did not give rise to modern humans. He had been hoping for a creature that looked more like Sir Arthur Keith's ideal ancestor—one with a bigger brain and more modern teeth that would clearly fit in the genus *Homo*. (The definition of the genus *Homo* at the time required a brain of at least 750 cubic centimeters, which was larger than a gorilla's brain but less than the smallest known human brain.) He called those "true men," and the australopithecines "near-men." But his disappointment gave way to excitement as he saw other humanlike characteristics in the skull. As he and Mary excavated the skull, it became clear that it was a robust australopithecine with enormous jaws and teeth, which would prompt one colleague to dub it "Nutcracker man." Interestingly, Louis pronounced Zinj a hominid on the basis of traits in its skull and teeth—not on any direct evidence from the skeleton for upright walking.

Louis called it "the connecting link between the South African near-men and true man as we know him." The robust skull is recognized today as a member of *Australopithecus boisei* (also known as *Paranthropus boisei*). More members of the same species, including a partial skeleton, have been found in Ethiopia and Kenya, and today *A. boisei* is considered more of a cousin whose own lineage went extinct than a direct ancestor of humans. Along with the australopithecines in South Africa, Zinj reinforced the hypothesis that Africa was the birthplace of humanity.

The discoveries also offered a new view of human evolution, showing that the earliest sign of becoming human was not a big brain, as Sir Arthur Keith had predicted (and, interestingly, as Louis had believed, as well). The Taung baby's species and Zinj, as well as more robust forms of australopithecines from South Africa, showed that one of the first steps toward becoming human was walking upright. The big brain did not come until later, less than 2 million years ago, when it began to expand in *Homo erectus*, reaching its largest size in Neandertals and modern humans.

The recognition of Africa as the birthplace of humans came at the same time as a revolution in the dating of fossil sites in the 1950s and 1960s. Louis had claimed that Zinj lived more than 600,000 years ago, and geologists thought at the time that the Pleistocene—the Age of Man—was about 1 million years ago (that date of 1 million years had doubled since the 1920s). Thus, Louis and Mary were stunned when, in 1961, a team of geologists at the University of California in Berkeley announced a date that was much older.

Berkeley geologists Jack Evernden and Garniss Curtis had traveled to Olduvai in 1957 and 1959 and collected samples of dirt at the base of the layer of sediments where the skull of Zinj was found. Researchers are unable to date ancient fossils directly, since they lack the radioactive elements necessary for radiometric dating. Instead, geochronologists collect samples of sediment laid down directly above and below a fossil, if possible, to bracket the age of the fossil. Back in their lab at Berkeley, Evernden and Curtis applied a sophisticated new method they had helped develop with physicists at Berkeley in the 1950s to the sediments from Olduvai. The method, known as potassium-argon dating, gave the age of sediments by using the radioactive decay of elements found naturally in rocks and soil. The method capitalized on a century of research that showed that unstable elements in rocks, such as radioactive uranium, potassium, and argon, decay gradually over time, emitting radiation and producing daughter elements that are stable and, therefore, not radioactive. The decay happens at a constant rate: the unstable version, or isotope, of potassium in volcanic sediments—called potassium-40—steadily decays, or ticks over, into a stable isotope of gas—argon-40—at a known rate.

In their lab, Curtis and Evernden heated the samples of sediment until they emitted the gas trapped inside. The pair used sophisticated new instruments to isolate and count atoms of potassium and argon gas as they were released from the sediments. The ratio of potassium-40 to argon-40 told them precisely how old the sediments were—the more argon, the more time had passed since the sediment was laid down at the fossil site.

The same methodology is also used with other isotopes, but the preferred method at ancient fossil sites in Africa today is a new version of the potassium-argon dating where researchers study the decay of two isotopes of argon. They use lasers to heat and count individual crystals of the isotopes of argon-40 and argon-39. The quality of the dates, however, depends on the quality of the sample of volcanic sediments, which must be rich in potassium or argon. Where fossils are found in soil without volcanic crystals, such as in Chad, researchers use other, less precise methods—such as dating sediments by the identification of extinct animals whose age is known.

In 1961, Curtis and Evernden were the first to use radiometric methods to date a fossil of an early human ancestor. Their potassium-argon dates for Zinj still hold; in this case, they dated the skull to about 1.75 million years. "It was four times older than previously thought at that time," said the eminent Berkeley anthropologist Sherwood Washburn. "It changed the conception of the rate of evolution regarding fossil man."

The new age fit Louis Leakey's notion that humans had been evolving in Africa far longer than previously believed. At about the same time, Evernden and Curtis also used the new method to date volcanic rock beds of known age in Italy that were a benchmark for the beginning of the Pleistocene. They showed, with others, that the Pleistocene had to be recalibrated to extend back to 2 million years, a date that still holds today.

For the Leakeys, nearly thirty years of work was justified, and fame unlike any they had known would soon follow. Louis embarked on a lecture tour of the United States and England, and entered into a long-standing arrangement with *National Geographic,* which started supporting the Leakeys' efforts, paying them salaries and buying them tents and equipment. They began serious excavations at Olduvai immediately—work that soon led to the discovery of fossils that fit the criteria for the genus *Homo.* Louis called them *Homo habilis,* or "handyman," because they were found alongside stone tools that he felt certain they had used. With this discovery, Leakey had a fossil that fit better his notions of an early human ancestor—and he proposed that *Homo habilis* was on the true line to humans, while all the australopithecines, including Zinj, were not. He held this view until his death in 1972.

But success brought problems at home. *"Zinjanthropus* had come into our lives," Mary wrote later. "Though we were not immediately aware of it, the whole nature of our research operation at Olduvai was about to be altered drastically, and we ourselves were going to be profoundly affected." The discovery of Zinj would have a "snowball effect" that would propel Louis and Mary down separate paths and eventually lead to the breakdown of their marriage. Where they had worked closely together before, Louis and Mary now traveled and worked separately more and more.

They were no longer alone, however, in their search for early humans in Africa. The French, who already had been working in Ethiopia, would expand their presence there, and a new generation of young researchers from Britain and the United States, unaccustomed to the ways of the British colonies, would push aggressively to get their own toehold on the prime fossil beds of Africa. For the first time, Louis would have serious competition in eastern Africa.

CHAPTER TWO

CONTINENTAL DIVIDE

The exceedingly cut-throat level of competition in Eastern African anthropology is a long-standing problem.

JEROME H. FREGEAU
former director of audit and oversight,
National Science Foundation, in a 1982 memo
to Vice President George H. W. Bush

On a searing hot afternoon in the summer of 1965, Bryan Patterson was walking back and forth across a gully, scanning the rocky desert floor. At fifty-four, Patterson was a seasoned paleontologist who had found dinosaurs in Colorado, early mammals in Texas and Wyoming, and extinct reptiles in Argentina. He had no college degree, but he had risen to the pinnacle of academia at Harvard University, where he held a prestigious chaired professorship in the Museum of Comparative Zoology. Now, on August 25, 1965, he was in the middle of his third summer in Kenya, and his graduate students were filming a scene that would become infamous in paleontological circles.

On film, Patterson is the epitome of the purposeful fossil hunter in the field, with a pick in his hand, poised to liberate a fossil from its sandy grave. As he walks, he is bent slightly at the waist, staring at the ground, like a beach-comber seeking the glint of sea glass among pebbles at the seashore. He is self-conscious, putting his hand to his brow as if in concentration. Suddenly, he swoops down over a gray bit of bone lodged in the sand. He lets out an excited cry and calls his team members to see what he had found. It is the classic eureka moment. Patterson had discovered a hominid, and he is captured in his moment of glory. Then the scene stops abruptly and the narrator, paleontologist William Sill, says: "That was Pat, showing the discovery like it ought to be."

A moment later, the jittery home movie starts again. Patterson is walking in the same gully, and Sill narrates: "This is the way it really was found." Now the retired curator of the Museum of Natural Science in San Juan, Argentina, Sill was a graduate student on that early expedition to Kenya with Patterson, and he helped film the discovery. This time, Patterson looks decidedly dazed. His shirt is unbuttoned; he's taking a drag from a cigarette. He looks as though he is about to wilt. "The afternoon temperature is 120 degrees. Your brain is fried. Pat's partially out of his mind from heat and on automatic pilot," says Sill. Patterson wanders almost aimlessly back and forth across the gully. Then he stoops to look at a bit of bone on the ground. He casually chisels the bone out of the sandstone with a pick, then puts it in his pocket and mentions to a team member that he has found a "knucklebone." Later he would say that he thought at the time: "Ho hum, there's another knucklebone." He walks on.

Almost fifteen minutes pass, Sill tells us, before the image of the bone finally penetrates Patterson's heat-addled brain and he takes the fossil out of his pocket and shouts, "My God, it's a hominid!" On film, he runs back to the spot where he found it. He tells other members of the team that it might be a bit of bone from an early human, and they all begin to search in earnest for more fossils. But they sifted through the sand in vain—they never found another hominid bone among the fossils littering the floor of this valley.

Later that week, Patterson's team raided a grave of the local Turkana tribe to compare the fossil with the elbow of a modern human—a practice that is obviously taboo today and could get a foreign team thrown out of a country. But in 1965, no such worries plagued Patterson, who did not want to wait for weeks until he could return to Nairobi to compare his fossil with human and ape bones in the National Museums of Kenya collection. One of his former students, paleontologist Roger C. Wood, recalled that the team needed an arm bone of a modern human and "none of us was willing to undergo an amputation." The team found what it wanted—a humerus. When Patterson aligned his ancient fossil alongside the humerus from a modern Turkana, he had his answer: his fossil was probably the elbow bone of an ancient human, as he had suspected. Later, he would compare the humerus with those of chimpanzees to make sure that it was more like that of a human than an ape. Eventually, scientists were able to pinpoint the fossil's age by getting dates from the radioactive isotopes in soil where the fossil was buried. At more than 2.5 million years old, it turned out to be the "oldest known manlike fossil," according to a report on the discovery in 1967 in the *New York Times*.

It came from Kanapoi. Patterson found the elbow bone in the same valley where Meave Leakey and the Hominid Gang would return almost thirty years later, to discover their own set of spectacular fossils. Meave Leakey's team also would redate the sediments that included Patterson's fossil and find that the real age of the elbow was about 4 million years—and that this hominid was the first known from the Pliocene epoch, 2 million to 5.3 million years ago. Kanapoi and the fossil fields on the western side of Lake Turkana would eventually turn out to be one of the epicenters in Africa for fossils of hominids, where the remains of some of the earliest members of the human family would emerge to shake up the human family tree.

But when Patterson, Sill, and other team members set eyes on the large valley of Kanapoi on July 7, 1965, they were not searching for hominids. They were seeking ancient mammals. They could not believe their luck when they came over a ridge and spied Kanapoi, which was on no maps. They saw so many fragmentary fossils of ancient animals strewn across the hills and gullies of Kanapoi that Sill wrote in his journal that the team felt as if they had found the ruins of an ancient Greek temple. "Our new find is unbelievable," he wrote. "There is bone everywhere. This is how the American west must have been before fossil hunters had picked over everything."

Patterson also realized that the fossils in the ancient valley were older than those found by Louis Leakey in Olduvai Gorge. The ancient age of the valley gave the team an added incentive to scour this valley for every kind of fossil, even keeping an eye open for the human kind. "Hooray, we found out by comparison of some of our fossils with Leakey's stuff that our site is just earlier than Leakey's oldest stuff," Sill wrote in his journal on July 21. "If we can only find a hominid, our little valley will be one of the most important places in the world (at least to paleontologists)."

After Patterson finished that season at Kanapoi, he returned to Nairobi and showed the fossil to Louis Leakey. Leakey's half-joking response was: "But you were not supposed to find hominids!"

Louis Leakey had invited Patterson to Kenya, along with the brilliant paleontologist George Gaylord Simpson from the American Museum of Natural History, who founded the Society for Vertebrate Paleontologists, to search for fossils of the ancestors of the wildlife living on the savannas of eastern Africa. Simpson passed, but Patterson accepted the invitation and checked in with Louis when he arrived in Nairobi in 1963, as was the custom. Louis welcomed him by sending him on a detour, steering him to three different places where the fossil beds proved a real disappointment. Patterson found almost nothing in those fossil beds during the long, hot summer of 1963. When he complained about those sites to Louis later, Louis answered that he was not surprised because he had never found anything there either.

The Leakeys' discovery of Zinj in 1959 and *Homo habilis* soon after, in 1961, triggered a new wave of American and European paleontologists eager to sweep out over the fossil beds of eastern Africa and find fossils of their own. A new team came to eastern Africa each year for the next twelve years after their discovery of Zinj, and many of these teams excavated for many seasons. After forty years of staking out promising fossil sites and toiling in the trenches without adequate funding or support from foreign colleagues, Leakey felt entitled to reserve the most promising areas for finding fossil hominids for himself—or for his chosen associates. He could be quite generous, particularly with young people who shared his passion for studying apes or fossils of early human ancestors. He is best known as the mentor to the three "ape ladies," as they were sometimes called: Jane Goodall, who studied chimpanzees at Gombe; Dian Fossey, who studied gorillas in Rwanda and Zaire; and Birute Galdikas, who studied orangutans in Borneo. He also invited many foreign scientists to work in Kenya and Tanzania. Nonetheless, he was notorious for being territorial, and he responded to the growing competition to work in the fossil fields of eastern Africa by dividing up the turf. "One woman in England had hippos, another Dutch researcher had horses and rhinos, a guy in Israel had crocodiles," recalls Patterson's former student Wood, a paleontologist at Stockton State College in New Jersey. The unspoken rule was that if they found hominids or fossil apes, they were to call in the Leakeys.

Leakey's tactics bothered Patterson, but they did not deter him. He was a self-educated man who had overcome many obstacles in his scientific career. He had earned only the equivalent of a high school degree from his preparatory school in England. But through sheer love of fossils and perseverance, he had become one of the world's leading experts on the evolution of mammals—first as a curator at the Field Museum of Natural History in Chicago, where he had gotten his start in paleontology in 1926, and later as a senior scientist at Harvard. At the age of seventeen, Patterson had walked into the director's office of the Field Museum and announced that he was ready to go to work. He had traveled to Chicago from his native England at the urging of his father, Colonel John Henry Patterson. Colonel Patterson was well known for slaying the famous man-eating lions of Tsavo, Kenya, that had killed more than a hundred workers who were building a railroad bridge—an adventure that Patterson had described in his 1907 best seller, *The Man-Eaters of Tsavo*. By 1926 he was more concerned with a restless son than wild lions, and he was looking for a way to help the teenage Patterson find a way to focus his energy. Colonel Patterson knew the director of the Field Museum, where the preserved Tsavo lions still are on display, and asked him if he could find a job for his son as a fossil technician. The father's introduction proved farsighted—young Patterson was indeed intrigued by fossils and learned fast, spending hours in the library teaching himself about mammal anatomy and taking courses at the University of Chicago. Eventually, he rose through the ranks to become a curator.

By the time he got to Kenya, Patterson held the prestigious Alexander Agassiz Professorship in the Museum of Comparative Zoology at Harvard and was a leading authority on the evolution of mammals of North America. He had also led many expeditions in North and South America. So when Louis Leakey sent him in the wrong direction the first season he was in Kenya, Patterson took it in stride.

He returned the next year, this time getting Louis Leakey's consent to prospect new terrain in the desert on the western shore of Lake Turkana, as well as assistance from Louis in coordinating the expedition. French and British geologists had surveyed the area in the early 1930s, and Leakey had explored the region a bit. Patterson knew there were deposits there that were at least 3 million years old and that spanned back through the beginning of the Pliocene and into the Miocene epoch (5.3 million to 23.8 million years ago). The boundary between the early Pliocene and the late Miocene, from 3 million to 7 million years ago, was a key moment in mammalian evolution—a diverse range of mammals were evolving in Africa and many mammals had not yet disappeared in mass extinctions. It would also turn out to be a crucial time in human evolution, although scientists did not yet understand how critical this slice of time was for humans. In the early 1960s, it was just one of the many gaps in the evolutionary road from early apes to *Homo sapiens.*

When Patterson first explored west Turkana, it was a particularly wild and inaccessible part of Kenya. The Turkana people, whose population numbered about 150,000 at the time, had been the last tribe to be brought under the control of the British army when Kenya was still a colony of Great Britain. In 1964, when Patterson's team first headed to Turkana, Kenyans had just won freedom from Great Britain—an independence movement known as Uhuru, which included the bloody Mau Mau rebellion to oust Europeans from Kenya—and the young nation was still in turmoil, with different tribes jockeying for power. Patterson was warned that the Turkana men were primitive, fierce warriors. But Patterson's real obstacles didn't turn out to be the

Turkana warriors, who proved friendly and whose spears look quaint next to today's weapon of choice, the AK-47.

Patterson's challenges proved much more mundane—his team's movies show a Land Rover lost in a river; Patterson himself crossing a rain-swollen river by dangling from a pulley as it moves along a cable over fast-moving waters; a line of men slowly pulling a truck by rope across a sandy riverbed. Patterson also had a reputation for getting lost. Tall and balding, with bony knees protruding from his khaki shorts, Patterson was the antithesis of the stereotypical bronzed explorer. He was a chain-smoker who read by kerosene lamp late into the night and slept in almost every morning, even in the field. He was notorious for "being lethal when left unsupervised in charge of machinery," says Wood. The team once spent almost an hour getting Patterson's foot out of a tire pump. He had no sense of direction, and his team members quickly learned never to let him drive if they wanted to get to their planned destination.

Patterson didn't take himself too seriously, but he was passionate about fossils—and clearly knew his way around fossil fields. Only two years after he found the fossil elbow at Kanapoi, his team discovered an ancient jaw-bone, this time in the red rocks of the Lothagam valley, north of Kanapoi, in the summer of 1967. Soon after Patterson's team discovered Lothagam, he learned that the sediments—and probably the fossil hominid—were at least 5 million years old. More homemade movies show Patterson, pate covered with a red bandanna, sitting at a camp table looking at the jawbone. It was the partial jaw of an ancient hominid that he considered an australopithecine, though he resisted the temptation to name it a new species. This weathered scrap of jaw and the elbow from Kanapoi would endure for thirty years as the benchmarks for the oldest known putative hominids. Yet Patterson's timing was too early—many years would go by before paleoanthropologists could grasp the full significance of these isolated fossils, since they still were debating when the earliest hominids would have been alive.

<hr />

The same summer that Patterson found the jaw fragment at Lothagam, another American was finally fulfilling a decade-long dream to search for fossils in the Omo River valley of southern Ethiopia, just 200 kilometers to the north of Lothagam. F. Clark Howell was a paleoanthropologist at the University of Chicago, and he had taken a course that Patterson taught. He had also won Louis Leakey's trust and friendship at a time when such backing could open doors—and even borders between nations.

Howell had first met Leakey at the end of a grand paleontological tour of Africa in 1954, while he was working on his Ph.D. in anthropology at the University of Chicago. He had grown up on a farm in Kansas and had a self-sufficient nature, which was honed by three years in the navy during World War II. Louis took a liking to the young man from Kansas and agreed to help him in 1959 when Howell wanted to survey the remote lands across the Kenyan border in southern Ethiopia, particularly along the Omo River, which flows into the northern end of Lake Turkana.

Howell had met the French paleontologist Camille Arambourg, who had found fossils in the Omo River valley twenty-five years earlier, in 1934. Arambourg's eyes had twinkled when Howell asked if he could join Arambourg should he ever return to Ethiopia. But Arambourg had many interests, so Howell eventually decided to take matters into his own hands. With Louis's help, Howell raised money, bought a used Land Rover, got a permit from the Ethiopian embassy in Washington, D.C., and headed north. But when he arrived at the border of Ethiopia, he was refused entry. After several days of hanging around the police camp at the border, he was finally allowed to enter Ethiopia, with police officers in tow. Howell patiently explained the geology to the police officers and showed them how to search for fossils. But when it was time to leave, the colonel in charge confiscated the fossils. "I tried everything to get the fossils out that I had collected," recalled Howell. "But more importantly, I could see the potential of the Omo." He could see that the layers of fossil-bearing sediments represented more time than Arambourg had realized—and geologist Frank Brown, of the University of Utah, would later confirm that they were between 1.8 million years and 2.5 million years old.

Howell returned to Nairobi and enthusiastically reported the potential of the Omo to the Leakeys over dinner in August 1959. After the dinner dishes were cleared, Louis took out a metal cookie tin and offered Howell something sweet. "I have a surprise for you," he said. "Have some dessert." Howell opened the tin and found the skull of *Zinjanthropus* inside. It was just two weeks after the Leakeys had found the fossil that was their first significant skull of a hominid, and they had not yet announced its discovery. Louis had singled out Howell as the first paleoanthropologist who wasn't a member of the Leakey family to see it.

Louis would remember Howell's enthusiasm about the Omo a few years later when he finally met with Ethiopia's Emperor Haile Selassie during a luncheon in Nairobi hosted by Kenyan president Jomo Kenyatta. In a now-famous exchange, Haile Selassie asked Leakey, "Why has my country got no fossils like you find in Tanzania and in Kenya?" Leakey recognized a good opening when he got one. He replied, "Well, your Royal Highness, if you would allow us to go and search in your country, I know where we might find something." Soon after, Leakey got permission to organize an expedition. He sent a cable to Howell saying simply: OMO OK. SEE YOU SOON.

On June 3, 1967, a convoy of three trucks, eight Land Rovers, and some forty people left Nairobi and drove five hundred miles north, into southern Ethiopia. Once there, the teams of researchers, from the United States, France, and Kenya, set up camps along the crocodile-infested Omo River. It was promoted as the first cooperative international expedition. In truth, it started out as three separate teams, each with their own leaders, financing, and unspoken desire to be the first to find the oldest, most spectacular fossils of human ancestors. Howell, then forty-two, was head of the American team; Arambourg, then eighty-two, led the French and Ethiopians (although he soon ceded leadership to French paleoanthropologist Yves Coppens); and Louis's son Richard Leakey, then twenty-three, was team leader of the Kenyans. Louis Leakey, Arambourg, and Howell had flown over the region and selected the outcrops of exposed rock and sediment where their teams would work. But once on the ground, only the French were pleased. Their section contained the oldest sediments, which were 2.5 million years old. Howell and Leakey both recognized that the sections they had drawn were far younger than those where the French were working. Howell was grumpy about it until the geologist on his team, Frank Brown, stood atop his Land Rover and spied a hill with exposed sediments that looked even older. Brown, then a graduate student at the University of California at Berkeley, would later confirm that the sediments were indeed 3 million years old.

Richard Leakey was not happy with his lot either. Although his team found the first fossil of the summer—fragments of two skulls of recent humans that lived about 100,000 years ago—he wanted to find older fossils. He was also eager to break out on his own, rather than work on an expedition his father had organized. The perfect opportunity came two months later when, flying back from Ethiopia in a light airplane, he spotted eroded sediments on the east side of Lake Turkana. When he returned, he found stone tools and fossils that he thought were older and more complete than anything he had seen in the Omo valley. He made a decision to shift his attention to Koobi Fora as soon as possible instead of returning the next year to the Omo. The decision was prescient—at Koobi Fora during the late 1960s and early 1970s, Leakey would find a spectacular series of fossils of australopithecines and early members of the genus *Homo* that would change paleoanthropologists' view of this key time in human evolution.

The French and American teams worked in the Omo valley until 1974, and found fossils of hominids at eighty-six different sites. Most were teeth or damaged fragments that were not as spectacular as the skulls at Koobi Fora. The Omo was a place of ancient rivers, so many of the fossils had been tossed and tumbled in fast-moving water. Only the hardest fragments—teeth and sturdy pieces of bone—were left in the sediments of the Omo. *National Geographic* photographers did not rush to the Omo to photograph the fragments, nor did Howell end up on the cover of *Time*. But to paleoanthropologists, the Omo Expedition would leave an important legacy—it would set a new standard for paleoanthropological expeditions and raise fossil hunting to the level of science (even the term "paleoanthropology" was new—it was first used in the 1950s). Earlier generations of fossil hunters had focused on collecting fossils of hominids almost as if they were on treasure hunts, picking up specimens randomly. They collected the prized hominids, but left behind other fossils of animals and plants that were full of information about the world where these ancient hominids lived. Raymond Dart, for example, never went to see where in a South African limeworks the Taung baby skull was found. Louis Leakey had learned the hard way when his iron stakes were stolen at Kanam in the 1930s that it was not enough to have the fossils—he had to be able to show precisely what layer of soil they came from.

F. Clark Howell was the first of a new generation of paleoanthropologists to insist on putting the fossils into geological context. He meticulously excavated each fossil by digging around it, leaving it on a pyramid of sediment, like a bone offering atop a pedestal. Geologists such as Frank Brown collected samples from the layer of sediment above and below the fossil for new state-of-the-art radiometric dating. They opened test trenches to see where more fossils were concentrated, and used surveyor's transits to map the horizontal and vertical positions of fossils on a map of the locality. Paleontologists who specialized in recognizing different types of animals, ranging from tiny rodents to large antelopes and pigs, recorded almost fifty thousand specimens, representing 140 species of mammals, and entered them into the first computer log of fossil mammals. One paleontologist, Gerry Eck, developed the first systematic method for keeping track of the types of animal fossils collected at each fossil locality. He also developed rules for what bones should be collected as a representative sample—how many of the abundant hippo teeth or antelope leg bones, for example.

The point of all this painstaking work was to situate the hominids in time and in their ancient space. By the time paleoanthropologists found fossils, the landscape had changed dramatically from the lost worlds these ancestors had inhabited. Their support crew's task, like detectives, was to collect evidence at the scene of an ancestor's death to reconstruct the setting where it had died. Fossils of tiny micromammals, such as mice and shrews, offer clues because they do not range far in their lives and are sensitive to environmental changes. Certain species are known to have lived in the desert, while others lived in wetter conditions. Hippos and crocodiles are obvious indicators that the terrain, now a dusty desert, was in ancient times swampy or near a river or lakeshore; while grazing antelopes, horses, and giraffes suggest grasslands.

The Omo Expedition was the first major expedition to reflect the growing sophistication of multidisciplinary paleoanthropological research. It marked the first time that a team of international researchers trained in so many different scientific disciplines would come together to work on one project, and it set the standards for large-scale paleoanthropological research in Africa. This is standard practice today. But it was an entirely new approach in 1967.

That year, 1967, was a remarkable one for paleoanthropology. It had begun with the press conference at Harvard's Museum of Comparative Zoology on January 13, 1967, where Bryan Patterson announced his discovery of a hominid from Kanapoi—the fossil elbow. William Sill's "little valley of Kanapoi" had indeed become one of the most important places in the world for paleoanthropologists. Later that year, Patterson's team would find the even older lower jaw from Lothagam whose identity would mystify paleoanthropologists for three decades. At the same time, the Omo Expedition was two hundred kilometers to the north, digging into the 2.5-million-year-old sediments in southern Ethiopia. And across the lake from Patterson, Richard Leakey would collect his first fossils at Koobi Fora, on the eastern shore of Lake Turkana.

But even as paleoanthropology was becoming a boom business in eastern Africa, Patterson's days working in Kenya were numbered. The new sophistication also meant new rules, and a new order in Kenya, where Richard Leakey was taking over administrative responsibilities from his father, Louis, at the National Museums. When Patterson wrote Louis Leakey in the spring of 1968 to tell him he planned to return to work in Kenya that summer, he got a letter from Louis saying it was no longer sufficient to declare that he wanted to work in Kenya. Instead, Patterson would have to apply for a permit with the Ministry of Natural Resources, which had tightened its controls over scientific expeditions. When Patterson asked for clarification, Richard Leakey responded that the new regulations were designed "to ensure that the work is in fact scientifically conducted." Patterson asked for further clarification.

Richard explained in another letter that there had been occasions when sites had been ruined and the fossils had been collected without proper excavations or records kept. Specifically, he placed blame, saying that the "worst offenders are Geologists." Richard had visited Kanapoi when Patterson was there and had complained that there were vehicle tracks over the fossil beds.

Patterson, whose small team was made up primarily of geologists, was so offended that he told Louis he would not be working in Kenya that summer. He never returned. Interestingly, in 1979 Patterson's obituary would emphasize his studies of fossil mammals and his significant contributions to understanding the evolution of mammals. But his fellow paleontologists who wrote the obituaries omitted reference to one type of mammal—the hominids from Kanapoi and Lothagam, perhaps because the search for human ancestors in west Turkana was unfinished business for Patterson. Remarkably, Kanapoi would remain virtually unexplored for almost thirty years.

CHAPTER THREE

THE EARLY ANCESTOR

Where, then, must we look for primaeval Man? Was the oldest *Homo sapiens* Pliocene or Miocene, or yet more ancient?

THOMAS HENRY HUXLEY, 1863

On January 14, 1967, Patterson's fossil elbow was featured on the front page of the *New York Times* as "the oldest known manlike fossil." At the same time New Yorkers were opening their morning newspapers and getting their first glimpse of their ancient ancestor's elbow, Louis Leakey was holding his own press conference in the National Museum in Nairobi, where it was already midday. At sixty-three, Leakey was still a showman who loved to surprise his colleagues. Leakey's bombshell was an even older human ancestor. The fossils he displayed had belonged to hominids that had been alive in Kenya 19 million years ago. It was a real jaw-dropper.

At the press conference, the white-haired Leakey held up the teeth and jawbones from eight adults and one infant and proclaimed that they were "the oldest so far clearly identifiable remains of Hominidae—the family of man." They were members of the genus he called *Kenyapithecus* (or "ape from Kenya"). "Man's separation from his closest cousins—the apes—is now carried back more than a million generations," Louis claimed. If he was correct in supposing they were early hominids rather than extinct apes, these fossils pushed back the time when the ancestor of humans split from the ancestor of African apes to at least 20 million years.

Leakey admitted that the fragments of jaws and isolated teeth did not look very spectacular "to the untrained eye." But to his eye they revealed a whisper of humanness—low broad crowns on the teeth and a rounded chin not seen in living apes. At this early age, a protoape would look a lot like a protohuman. There were no clear-cut criteria for recognizing a species of hominid this ancient, because most of the obvious hallmarks of humanness, such as upright walking, language, and a big brain, had not developed yet. On the basis of the assortment of teeth and jaw fragments—some found years earlier—he proclaimed these new fossils the earliest members of the human family, and direct ancestors to the true hominids that appeared in Africa much later.

The press featured Leakey's claims uncritically, even though many of his scientific colleagues were skeptical. The headline in the *New York Times* the following day, January 15, 1967, announced: MAN'S ANCESTOR FOUND TO BE MORE THAN 19 MILLION YEARS OLD. The London *Times* noted that "Dr. Leakey's claims are potentially much more important for the study of man's ancestry than the discovery reported from Harvard." It probably was not lost on Leakey that Patterson and the Harvard team had enjoyed the limelight for only one day.

The gap between Leakey's and Patterson's fossils from Kenya was 16.5 million years. A reader of the *New York Times* the day after Leakey's press conference might have wondered what the distinction was between "the oldest manlike fossil" at 2.5 million years and "the oldest known ancestors of man" at 19 million years. Which one of these fossils from Kenya was the first member of the human family?

The juxtaposition of those two headlines vividly illustrated a fundamental problem before human paleontologists in the 1960s and early 1970s. Did the first member of the human family arise early—more than 14 million years ago, in the Miocene (the epoch that ran from 5.3 million to 23.8 million years ago)? Or was it a relative latecomer on the scene in Africa, first walking upright 5 million years ago or so at the end of the Miocene or just after in the beginning of the Pliocene (the epoch that ran 2 million to 5.3 million years ago)? Patterson's fossils were the first of a parade of bona fide hominids to appear in Africa and Asia at that late time. But it was unclear whether these African hominids were late revisions of a human form that had begun taking shape 10 to 20 million years ago. Or were they the earliest drafts of protohumans that had only lately crossed the subtle line from being an extinct ape to being an early human ancestor?

Almost all paleoanthropologists in the 1960s had been taught that the ancestors of humans split from the ancestor of the great apes in the middle to early Miocene, then newly dated to between 10 and 20 million years ago. If Leakey was right that *Kenyapithecus* was a hominid, it was alive when many researchers expected that the ancestor of humans had already separated from the ancestor of apes. Indeed, the hypothesis that the human lineage was extremely ancient was an old idea with widespread support. Charles Darwin had written a century earlier in *The Descent of Man* that man must have had a pedigree of "prodigious length," because humans had undergone a great amount of modification in comparison with the higher apes—and he speculated that the great apes had diverged from one another during the Miocene.

Darwin had been followed by a procession of paleontologists whose estimates of the timing of the separation of the human lineage from other primates ranged from 14 million to 40 million years ago. One of the most extreme estimates came from Henry Fairfield Osborn, a well-known paleontologist at the American Museum of Natural History in New York City, who pressed the view that humans had split off from primates in the Oligocene epoch (now dated 26 to 37 million years ago). This notion of humans' early origins was based on the assumption that the great apes were more closely related to one another than to humans. This view allowed enough time for humans to separate from the ancestor of all the apes, and then for that protoape to give rise to the diverse forms of chimpanzees, gorillas, and orangutans. It also had the virtue of allowing plenty of time for the special, human lineage to produce the highly evolved traits thought to be unique to humans—the big brain, upright walking, hairlessness, tool use, and language.

The fossil evidence for this view was scanty at best. Although there had been a virtual population explosion of hominids in the fossil record in the century since the discovery of the first hominid fossil in the Neander valley in 1856, there were still so many gaps in the chronology that it was risky to use fossils to prove that the human lineage had started so early. At the time, more than one paleoanthropologist warned that all of the known fossils of early humans could fit on a billiard table. But that didn't stop many paleoanthropologists from trying to connect the dots between the fossils in the Miocene to recent African fossils and right down to modern humans. Louis Leakey's attempt to link his 19-million-year-old fossils to the "true" hominids that lived 2 million years ago was just the most brazen example: in Leakey's view, Miocene hominids, such as *Kenyapithecus,* were ancestral to the fossils of the earliest members of the genus *Homo* that his team had found in Olduvai. In his view, the australopithecines that lived in Africa, such as the Taung baby and Patterson's fossils, were not direct ancestors to modern humans. Even though Louis recognized the australopithecines as hominids, he still clung to the notion that human ancestors had big brains—and the small-brained australopithecines belonged on some side branch, as distant cousins of the true ancestor.

But even Leakey could not ignore the gaps in the fossil record. In the foreword of his last book, *By the Evidence,* his editors acknowledged that the fossil trail vanished between 14 million years and 2 million years ago. Even where the trail was most complete—in Africa in the past 2 million years—it was still hard to follow.

All of these new discoveries had put Africa firmly on the map as the birthplace of the genus *Homo*—and, therefore, of true human ancestors. Many researchers, unlike Leakey, had also accepted at least one type of australopithecine as ancestral to modern humans. But the immediate ancestors of the australopithecines were missing. Patterson's elbow and jawbone were the only two fossils representing the vast span between 2.5 million and 5 million years ago. They were the solitary remnants of populations of missing hominids, like lone sentinel stars marking the presence of invisible galaxies too distant to detect.

The next group of fossils to appear on the horizon were so ancient that they were like aliens from a distant planet. It was impossible to tell if any of these apes from the middle Miocene, 11 million to 16 million years ago, were the ancestors of the hominids that arose much later in Africa. The first apes appeared in Africa between 17 million and 23 million years ago, and though there is debate about which fossil represents the first "ape," they came in several forms, including *Proconsul africanus* at 17 to 19 million years and Leakey's *Kenyapithecus* at 14 million years. By 16 million years ago, apes had spread into Europe and Asia, where the climate was warmer than it is today. By the 1960s, so many fossils of Miocene apes had been discovered that it was clear the earth truly was the planet of the apes, at least during the middle Miocene. They came in all sorts of shapes and sizes, each one seemingly with a new name: *Dryopithecus, Oreopithecus, Ramapithecus, Sivapithecus, Kenyapithecus, Pliopithecus, Bramapithecus,* and so on. They flourished during a time when the Tethys Sea, which had long separated Africa and Eurasia, was retreating and creating a giant landmass thick with rain forests. Gangs of apes moved between the continents, traveling along thickly forested corridors from Spain to China.

But the golden age of apes came to an end when almost all of these apes went extinct, disappearing first from the fossil record in Africa, then in Europe and Asia. They dropped off one by one as the climate cooled and got drier. The forests that provided them with fruit and sleeping nests were dwindling. The landscape was changing dramatically in Europe and Asia as giant continental plates squeezed together, pushing the Alps, the Urals, and the Himalayas to new heights. In Africa, tectonic forces were pulling apart the Great Rift Valley of eastern Africa, causing a flurry of volcanic eruptions. Forests were giving way to desert and more scrubby grasslands in eastern Africa. Many species of apes did not survive the remodeling of their habitats—and only a few persisted to give rise to modern humans and apes in Africa. The mystery of the Miocene was, Which one of these fossil apes was the one that survived to produce humankind? And did it arise in Africa, or was it part of a migration of Miocene apes that came back to Africa from Europe or Asia?

One young scientist who set out to solve the mystery of the Miocene in the 1960s was a young American named Elwyn Simons. Today, Simons is an eminent primate biologist at Duke University in Durham, North Carolina. In his mid-seventies, he looks every bit the individualistic sage that he has become: bearded, with glasses and, often, a surprising fossil of a new type of extinct primate up his sleeve. He is recognized as a formidable field-worker, known for his remarkable success in collecting fossils of extinct primates and for his studies of the behavior of living primates on some ninety expeditions to Egypt, Madagascar, India, Iran, Nepal, and Wyoming. He also is a master storyteller who retains a boyish enthusiasm for describing and collecting fossils of primates, including one infamous Miocene ape called *Ramapithecus*, whose identity confounded him and his colleagues for many years.

Simons' encounter with *Ramapithecus* began in 1960, when he was getting his start as a young researcher at Yale. When he arrived at Yale's Peabody Museum, he began working on a long-neglected collection of Miocene fossils found thirty years earlier in the Siwalik Hills of northern India by G. Edward Lewis, then a Yale graduate student. Lewis had found them in 1932 and named one of them *Ramapithecus brevirostris* (or "Rama's short-faced ape," named after the Hindu god Rama). He had caused a stir when he published a paper describing the jaw as more like a human's than any other fossil ape then known—and later claimed in his thesis that *Ramapithecus* belonged in the human family. But Lewis's suggestion that *Ramapithecus* was a member of the human family was shot down by one of the most influential paleoanthropologists of the time, in the same manner that Raymond Dart's Taung baby also was dismissed as an ape. Lewis eventually left Yale and never returned to work on *Ramapithecus* again.

By the time Simons arrived at Yale, the fossils of *Ramapithecus* had been virtually forgotten. Lewis's tools and materials were still at the museum, where they had been stored for Lewis, who would never return. As Simons read Lewis's thesis and studied the original fossils, he could understand what Lewis had seen in the jawbone. Was this fossil that had lived 8 to 9 million years ago an early hominid or an extinct ape? Simons delved headlong into the thorny problem of how to recognize an early hominid among the plethora of apes from the Miocene.

Simons had been well trained for solving this puzzle. He had two

Ph.D.s, one in paleontology from Princeton University, where he had studied extinct mammals, and one in human anatomy from Oxford University in England. He'd chosen Oxford because he'd wanted to study the origin of primates, and he'd soon found himself in the orbit of a constellation of particularly bright stars in paleoanthropology, including Sir Wilfrid Le Gros Clark.

Simons, who had grown up in Texas and was at Oxford on a Marshall scholarship, was a little awed by Le Gros Clark, who was twice his age and had been knighted by the queen for his contributions to science. Le Gros Clark was tall, almost regal in his bearing, and could be reticent, choosing his words carefully, perhaps because he had overcome a stuttering problem as a young man. Le Gros Clark was also in high demand and was revered by his students, who had commissioned a handsome bust of him, which sat outside his office. But Simons remembers that despite his fame, the older scientist was never pretentious. Le Gros Clark once took a look at the bust and commented to Simons, in his clipped British accent: "Well, if it's not the way I am, it's the way God should have made me."

One of the key questions that Le Gros Clark had worked on was how to organize the primate fossils of the Miocene. For generations, paleontologists had baptized each precious new fossil as a new species or even genus, with some species based on single teeth. Those researchers, such as Louis Leakey, who split two similar fossils into two species were known to their colleagues as "splitters," as opposed to "lumpers," who saw differences as variation within a species and lumped them into one species. Researchers were also keen to find traits in these extinct apes that might foreshadow a hominid in some way, and help them identify a protohuman among the protoapes.

Le Gros Clark, like most researchers, still thought that the most reliable hallmark of being human was walking upright. But there were no significant fossils of pelvic or lower-limb bones known then from any of these Miocene apes. As a practical matter, researchers had to find traits in the teeth and skull that set apart early members of the human family from apes.

Le Gros Clark had identified a short upper canine tooth that did not overlap—or rub—the bottom tooth as a sign of a hominid; indeed, many paleoanthropologists still regard a small, nonhoning canine in males as the mark of a member of the human family. A long, overlapping canine that rubbed against the bottom teeth when chewing was the sign of an ape. This short canine, Le Gros Clark thought, reflected a significant change in behavior, unique to the human family. Like most paleoanthropologists of the day, Le Gros Clark subscribed to an idea first proposed by Darwin that as human ancestors began to use tools, they no longer needed a huge, sharp canine for fighting or to display their strength to other males. Darwin also thought that the invention of tools came at the same time as upright walking and the big brain—by walking upright, early humans were able to use their hands to make, carry, and wield tools. A toolmaker required a bigger brain to create and use the new stone technologies. It was a neat package: the invention of a tool kit caused the first technological revolution, causing human ancestors to stand upright.

Le Gros Clark also looked at the roof of the mouth—the palate—and noted that it was arched in humans, instead of flat as it is in an ape's mouth. In humans, the teeth go back on each side in a broad, parabolic curve that is widest at the rear and in a V shape, instead of in the U shape of an ape, where the two rows of teeth are parallel and the distance between the back teeth is the same as that between the front molars. But even as Le Gros Clark listed these traits and others, he recognized that there was not one magic character—with the exception of bipedalism—that made a fossil a hominid instead of an ape. He emphasized instead the importance of a total pattern of characters that would set apart the ancestors of humans from the ancestors of great apes.

At Yale, as Simons looked at *Ramapithecus*, he tried to apply those criteria, but he only had scanty evidence—fragments of jaws and teeth. Yet he saw what Lewis saw: canines that were smaller than those in living apes, and molars that appeared squarish as in an australopithecine rather than an ape. The enamel was also thick, like that of an australopithecine. The complete palate was not preserved, but Simon's reconstruction showed rows of teeth in a parabolic curve resembling a human's, rather than the U shape in apes. All of this led Simons to pronounce *Ramapithecus* a possible hominid in 1961.

He soon was joined by David Pilbeam, who had earned his bachelor's degree at Cambridge University in his native England and was still in his early twenties. He had come to Yale on a fellowship in geology, with a deep interest in hominid origins and a recommendation from his adviser in England to look up Simons. He found him in the thick of working through the paleontological mess of the Miocene fossils. Pilbeam, who was soft-spoken and measured, soon rolled up his sleeves and began analyzing the fossils with him.

A clearing-out process was finally under way, with a vanguard of paleoanthropologists turning to biology to adopt the methods biologists were using to sort living species of animals to classify hominid fossils. The debate about how to name or classify a hominid fossil might seem purely academic. But this classification determines whether a fossil fits into the human family tree and if its species was on or close to the direct line to humans.

Traditionally, paleontologists had organized fossils into new species based on detailed descriptions of the differences in anatomy between individual specimens. This "typological" approach often meant that they used a "type specimen," or reference fossil, to define the species. They compared new fossils to the type specimen, measuring static characters, often with calipers, and assigning fossils to separate species based on the degree of difference in those traits. By contrast, since the 1940s biologists had been using an approach known as the New Systematics to classify organisms. The New Systematics was less of a method than an outlook, requiring biologists to think of a species as a dynamic population whose members showed variation in their anatomy—due to gender or differences in diet, for example—but who could interbreed and were part of a community.

Harvard zoologist Ernst Mayr, who had started his illustrious career by "cleaning up" the classification of birds of paradise in Papua New Guinea in 1928, had proposed this idea, calling it the "biological species concept." It essentially defines a species as a community whose members can interbreed successfully, and who share a particular habitat or adaptation that isolates them from reproducing with other populations, such as eating different kinds of food or using different strategies for selecting mates. One cannot define species or types by applying an arbitrary yardstick to differences in anatomy. It does not matter if two populations of birds looked exactly alike, such as dif-

ferent species of black-capped titmice or leaf warblers. If they cannot produce viable offspring, they belong to separate species. Conversely, if two
birds have beaks of different lengths but still produce offspring, they are
members of the same species.

While the New Systematics had swept through the fields of zoology,
botany, and genetics by the middle half of the twentieth century, hominid paleontology was one of the last holdouts of the old typological approach. For
paleontologists comparing fossils, it is virtually impossible to determine
whether specimens would have been capable of interbreeding—particularly if
the paleontologist has only parts of the skull or jawbone from one individual.
Instead, researchers have to sort through the traits they are comparing in the
fossils and decide which ones reflect real differences in adaptive behavior or
biology. Elwyn Simons, who had studied living primates, was conscious that
differences between two individuals might be due to their individuality, their
age, or their sex, for example. He and Pilbeam were lumpers at that time, and
they cast out many species as they reviewed the fossils.

One new genus of ape that Simons and Pilbeam eliminated with a
stroke of the pen was Leakey's *Kenyapithecus,* which at 14 million years was
older than *Ramapithecus,* which is now dated from 8 million to 13 million
years ago. Simons and Pilbeam came to the conclusion that they were members of the same species—and they called all the fossils *Ramapithecus,* because
under the international rules of nomenclature for naming new species, the
older name is kept. Leakey was not pleased with this conclusion.

The scientific community, already predisposed to the idea of an early
ancestor for the human lineage, embraced *Ramapithecus* as an early hominid
and the new simpler order for the Miocene apes. As they gathered more fossils of *Ramapithecus,* Simons and Pilbeam published a series of papers in
which they listed more than a dozen traits that linked *Ramapithecus* to australopithecines. These features suggested the radical notion that *Ramapithecus*
might have been using tools to prepare its food, verifying Darwin's ideas that
canine reduction came with tool use. "Hands were probably used extensively,
and perhaps tools as well," Pilbeam and Simons wrote in 1965. And if hands
were free to use tools, this suggested that perhaps *Ramapithecus* walked up-

right, although there were no fossils from beneath the neck—and all of this was based on fragments of jaws and teeth.

Thus, *Ramapithecus* took its place in the pantheon of earliest known human ancestors. It was shown walking upright across the pages of the Time-Life book *Early Man*. One popular textbook, *Human Evolution*, by Bernard Campbell, summed up the dominant view of *Ramapithecus* in 1966: "We can safely put this specimen near the origin of man's distinct lineage. On this basis we might suppose that the Hominidae became separated as a lineage at least fifteen million years ago."

CHAPTER FOUR

_____

DRAWING BLOODLINES

Discoveries made in a field by someone from another discipline will always be upsetting to the majority of those inside.

J. CRAIG VENTER, human genome researcher

What we know about human origins can be constrained by what we know about molecular evolution.

ALAN WALKER, paleontologist

Vincent Sarich was a graduate student at the University of California in Berkeley when he volunteered for an assignment that would change his life—and eventually the way researchers traced the pedigrees of humans and apes. It was 1964, and he was a graduate student in anthropology attending a weekly seminar taught by the eminent anthropologist Sherwood Washburn and his former student Clark Howell, who was a professor at the University of Chicago at the time and on the verge of organizing the Omo Expedition to Ethiopia.

At thirty, Sarich already made a vivid impression—at well over six feet tall with a shock of black hair, he was physically imposing. He was older than most of the other students because he had taken time off from school and had earned two undergraduate degrees (in chemistry and in anthropology), and he was not afraid to express his views. He had an uncanny instinct for honing in on controversial topics. Years later, he would tell a reporter for the journal *Science*, "Of course I like to raise hell"—in response to criticism that he had provoked his students to debate whether genetics and evolution might create differences in human behavior and intelligence between men and women and among people of different races.

In Washburn and Howell's seminar, Sarich was quickly drawn into the debate on one of the hottest topics of the day: whether *Ramapithecus* and other fossils provided compelling evidence to show that the ancestor of humans split early from the ancestor of apes. Washburn was one of the only prominent anthropologists of the day to openly reject the prevailing opinion that the ancestor of humans arose early in the Miocene era, more than 15 million years ago. He had studied anatomy and behavior in apes and thought that humans were sufficiently similar to the African apes that they had shared a long heritage, and had separated from a common ancestor only recently. Almost all of the really significant changes in humans had evolved in the past 2 million years, in his view, which meant that the separation took place in the past 4 million years or so, he thought.

Washburn also was one of the first biological anthropologists to recognize that it might be possible to use the new genetic technologies to test ideas on how close humans really were to different apes. Washburn had been influenced early in his academic career by the geneticist Theodosius Dobzhansky, who was a major figure in the development of modern genetic theory and who was on the faculty of Columbia University Medical School with Washburn in the 1940s. Although the use of genetic technologies to study primate evolution was a fledgling field, Washburn was keen to nurture their application to human evolution studies as an independent line of evidence from the fossils.

Sarich was in the right place at the right time. With his undergraduate degree in chemistry, Sarich quickly recognized that the question of when humans split from apes was an interdisciplinary problem that he was uniquely prepared to take on. He volunteered to write a review of the molecular data that had bearing on the evolutionary relationships of different species—including humans and apes. As Sarich got to work, he knew that most biologists thought that chimpanzees, gorillas, and orangutans were more closely related to each other than they were to humans, which meant that most researchers classified them together in the family Pongidae, a name that drew on an African Bantu-language word for ape. Humans were all alone in their own family, Hominidae (or Homininae), a name that is derived from the Latin word *homo,* for "man."

But this view of humans and their ancestors as the sole members of Hominidae had recently been challenged by Morris Goodman, a biochemist at Wayne State University's School of Medicine in Detroit. Goodman and a colleague had measured tiny differences in the surfaces of the same blood proteins in humans and apes. Goodman knew that work in the early twentieth century had shown that the structure of blood proteins became less alike as two species became less closely related. He decided to continue this line of work and went to the curator of the Chicago Zoological Park in 1957 to collect blood from chimpanzees, gorillas, orangutans, and gibbons.

Back in his lab, Goodman isolated a type of blood protein called albumin from a human and injected it into a living chicken. Goodman's test capitalized on the immune system's exquisite ability to defend itself from foreign invaders, such as viruses, by recognizing tiny differences on the surfaces of these molecules that reveal whether they belong or not—whether they are "self" or "not self." Once the immune system detects an invader, it goes on alert and synthesizes millions of different proteins, called antibodies. These antibodies then identify and attack a specific invader—called an antigen—by binding with it and destroying it.

In Goodman's initial tests, the chicken's immune system could not tell the difference between an albumin from a human, a gorilla, or a chimpanzee—and fought them with equal vigor. But it did not recognize the blood proteins from orangutans or gibbon apes as invaders, which meant they were less like the human blood protein. The results with several blood proteins supported the view that chimpanzees and gorillas were more closely related to humans than were orangutans and gibbons.

In the summer of 1962, Goodman excitedly presented his findings at a conference entitled Classification and Human Evolution held in the twelfth-century Burg Wartenstein castle in the Austrian Alps. Washburn, who had helped organize the gathering of about twenty leading researchers, heard Goodman's radical proposal. On the basis of his immunological data, Goodman had proposed that humans, as well as gorillas and chimpanzees, should all be put in the same family, Hominidae. He would leave only orangutans in the Pongidae and only gibbons in the Hylobatidae family. Goodman drew a family tree based on the immunological data—a phylogenetic tree, which shows the evolutionary relationships among a group of species. His tree showed humans, gorillas, and chimpanzees splitting from a common ancestor at the same time or close together, rather than humans veering off much earlier. "Naively, I anticipated that my proposal to so revise the classification of hominoids would be well received," Goodman wrote years later. Even though his version of the family tree was actually similar to those produced by Darwin and Thomas Henry Huxley in the 1860s and 1870s, the idea that humans should be put in the same family as chimpanzees and gorillas was controversial. Two of the most eminent scientists of the twentieth century—the paleontologist George Gaylord Simpson and the evolutionary biologist Ernst Mayr—disagreed. Although they recognized that humans were closely related to the African apes, Simpson argued that humans and their ancestors still should be classified in their own genus because of their unique adaptations.

The high-powered opposition to Goodman's proposal probably cemented Washburn's interest in taking on this problem. "Washburn and I were alike in that we both started worrying when people agreed with us," Sarich recalled. As Sarich surveyed the literature in preparation for his class in 1964, he realized that the next step would be to put dates on the branching points in Goodman's family tree—namely, when humans, chimpanzees, and gorillas shared a common ancestor. Although Goodman had shown chimpanzees, gorillas, and humans separating from a common ancestor at about the same time, he did not go further out on a limb and say when they parted company.

As Sarich thought about how to date the split more precisely, he learned about the pathbreaking work of Emile Zuckerkandl and the chemist Linus Pauling, who was famous for winning two Nobel Prizes and, later, for his advocacy of vitamin C. Working together at the California Institute of Technology, Zuckerkandl and Pauling had been the first to show that changes in molecules over time—specifically proteins—can be used as a clock to help date when two living creatures shared a common ancestor. The molecular clock relies on the notion that as species diverge from a shared ancestor, each separate lineage accumulates mutations in their DNA at a steady rate on average over long spans of time. Therefore, as more time passes, their DNA (and certain proteins they encode) becomes more different—and the number of differences is directly proportional to the time that has passed since two species separated from the common ancestor. Indeed, the molecular clock does not keep perfect time, but an overwhelming number of studies have since shown that it is generally reliable for sorting out how closely related many different types of animals are.

As Sarich prepared his seminar report in 1964, his flash of insight was that he could apply Zuckerkandl and Pauling's concept of a molecular clock to date when human ancestors split from apes. His proposal was a hit with Washburn, who would later introduce him to a young professor of biochemistry, Allan Wilson, who was a native of New Zealand and in his first year on the faculty at Berkeley. Wilson invited Sarich to join his lab, where they used Goodman's immunological method to sort the humans and apes on a family tree. They dated the branches with the molecular clock. They also tested the clock by inventing a rate test to make sure that mutations accumulated at about the same rate in different lineages of apes and humans. This meant that the number of differences in each lineage could be compared directly and, therefore, were generally reliable for calculating how long the branches between apes and humans on the family tree should be. Finally, they dated the age of all the branches in the tree by calibrating it with the most reliable dates for primate fossils. Before long, Sarich and Wilson had encouraging results, and Washburn sent a letter to Elwyn Simons at Yale telling him he had a secret weapon that would prove that apes and humans had diverged recently.

By the end of 1967, Sarich and Wilson had a date: the split of humans and the African apes was about 5 million years ago, not 15 million or 25 million years. When they were confident about their results, they published three papers in the course of a year on their methods and their date for the split. They concluded in the December 1, 1967, issue of the journal *Science* that "man and African apes shared a common ancestor 5 million years ago, that is, in the Pliocene era."

It was a molecular bombshell. When the article landed on the desks of paleoanthropologists, it created a decades-long rift between scientists who studied molecules and those who analyzed fossils. Louis Leakey would say that the date of separation suggested by Wilson and Sarich is "not in accord with the facts available today." Simons would protest that if Sarich and Wilson's dates were correct, then paleontologists had not found *a single fossil* that was ancestral to any living primate. "It is not presently acceptable that *Australopithecus* sprang full-blown five million years ago, as Minerva did from Jupiter, from the head of a chimpanzee or gorilla," Simons wrote.

The reaction was not a promising start for the introduction of the new molecular methods into paleoanthropology and paleontology. Paleoanthropologists were not ready for a paradigm shift in their view of how long it took hominids to evolve. The molecular dates had not been proven by the discovery of well-dated fossils, so there was no hard data to test the accuracy of the molecular clock. After new studies confirmed that humans had recently shared an ancestor with the African apes (using other proteins and new methods to compare the way segments of DNA, rather than proteins, bound with probes), Sarich would pronounce in 1971: "One no longer has the option of considering a fossil specimen older than about eight million years a hominid no matter what it looks like." He and Wilson not only pushed *Ramapithecus* out of the human family tree but also made it clear that they would oust any fossils from the human branch that were too old—even if they looked like human ancestors. Their conceit was clear: they obviously thought the molecular data should be taken more seriously than the fossil data for sorting out where different species or genera fit on the ape and human tree.

Many paleontologists and paleoanthropologists responded by ignoring or criticizing the notion of a molecular clock, although Washburn was an early champion of the method. But Sarich, now a professor emeritus at Berkeley, still holds to a date of between 4 million and 6 million years for the divergence of the human and chimpanzee lineages. Scores of studies with many different lineages of DNA in humans and apes have produced human family trees with chimpanzees on the branch closest to humans—a finding confirmed in 2005 by the publication of the draft sequence of the common chimpanzee genome. While there are small differences in the length of the branches between humans and chimpanzees—and therefore the estimate of the time that has passed since they separated from a common ancestor—all but one team's trees date the split of humans and chimpanzees to sometime between 5 million and 7 million years ago.

It would take more than a molecular clock to show that it was high time for *Ramapithecus* to come down from the human family tree. More than a decade would pass before many paleoanthropologists—including Simons and Pilbeam—would admit that *Ramapithecus* was no hominid. Even then, it would take the paleoanthropologists closest to the fossils to finish off *Ramapithecus* as a hominid—on the basis of its anatomy.

The beginning of the end for *Ramapithecus* came soon after David Pilbeam led an international expedition in 1973 to the same fossil beds where the Yale student G. Edward Lewis had found the original fossils in 1932. By then, Pilbeam knew he needed more than fossils of teeth and jaws to prove that *Ramapithecus* was on its way toward becoming a human. In a collaborative research project with the Geological Survey of Pakistan, Pilbeam's large, interdisciplinary team would work for more than three decades in the Pakistani section of the Siwalik Hills, which run along the southern flank of the Himalayan mountain range from Afghanistan to Burma. They would find thousands of fossils of animals; eventually, in 1975 and 1976, the team found jaws of *Ramapithecus*.

When Pilbeam fit together the pieces of the lower jaw, he could see a kind of parabolic U shape, not the V shape that Le Gros Clark had identified as belonging to hominids. Then the team found a partial skull and isolated limb bones of the more apelike *Sivapithecus*. When Pilbeam looked *Sivapithecus* in the face, he realized it was the male version of *Ramapithecus*—and that neither one was a hominid. The form of the face and the shape of its eye sockets, nose, and palate most closely resembled those of the sole surviving Asian great ape, the orangutan. *Ramapithecus* did indeed have smaller canines, but that was because it was female, not a hominid. By late 1979, both Pilbeam and Simons realized that *Ramapithecus* was just the female form of *Sivapithecus*, aptly named for the Hindu god of destruction.

Pilbeam discussed *Ramapithecus*'s fall from grace with good humor, although the episode clearly troubled him. In an article in *Scientific American* in 1984, Pilbeam posed the question this way: "Why was the hominoid fossil record misinterpreted by dimmer paleontologists such as me?" Pilbeam's answer was a warning to future anthropologists: "Too much attention has been paid to the fossil record as a source of information about evolutionary branching sequences." Fragmentary jaw and isolated teeth are parts of the body that are "seldom informative" for sorting out genera (although they are still useful for sorting species). *Sivapithecus* had shown, for example, that thick tooth enamel and small canines by themselves were poor markers of being a hominid. The discovery of thick tooth enamel in *Sivapithecus*, and subsequently in other extinct Miocene apes, proved that it reflected the animals' diet rather than recent common ancestry. A small canine alone also was not enough to prove that a fossil was a hominid, particularly if no fossils of males had been found.

Pilbeam, in fact, lost so much confidence in using fragmentary fossils of jaws and teeth for determining branching sequences that he became a key champion of molecular methods for arranging humans and the apes in the primate family tree. The evidence coming from molecular studies was becoming impossible to ignore. In 1982, two biologists at Yale had used a new method of studying DNA, rather than proteins, to show that humans were more closely related to chimpanzees than to gorillas (a result that has been confirmed many times, showing that humans and chimpanzees share about 98 percent of their DNA for many different segments of DNA). Pilbeam embraced the molecular data wholeheartedly. Indeed, where he had published family trees based on fossils in the past, he now published a tree in *Scientific American* based on blood-protein data from living primates. It showed the ancestor of gibbons splitting off first from the line leading to humans, more than 20 million years ago. Then the ancestors of orangutans split off at 16 million years; of gorillas at just less than 10 million years; of chimpanzees between 6 million and 7 million years ago; and, finally, australopithecines diverged from early *Homo* between 3 million and 4 million years ago—dates that still hold for most molecular anthropologists, depending on which fossils they use to calibrate their starting branch.

Pilbeam's enthusiastic acceptance of the molecular data troubled Simons and other paleoanthropologists, who grumbled that Pilbeam had gone too far in describing the limits of fossil data—Pilbeam was like someone who had lost his religion and converted to a new one. By then, Pilbeam and Simons were still friends but no longer a team at Yale; Simons had moved to Duke University in 1977 and Pilbeam to Harvard in 1981. Simons would eventually concede in *Science* that, on the basis of the hard data from the fossil record (not the molecular clock, of which he is still deeply skeptical), for the better part of two decades "searchers were looking in the wrong place, for the wrong ancestors, with the wrong anatomy, at the wrong time."

CHAPTER FIVE

LUCY, THE LATE ANCESTOR

Time that was so long grows short.

ANNE SEXTON

Not everyone was looking at the wrong time and in the wrong place for the earliest hominids. One young French geologist, Maurice Taieb, came across the right place at the right time—twice. But it was the late 1960s, and few people realized then how important a discovery he had made at the time, including Taieb himself.

Taieb was thirty years old and earning a Ph.D. in geology at the University of Paris when he first visited the Awash valley of Ethiopia in 1965. He was not the first geologist to explore the Awash; European explorers and geographers had been passing through the arid badlands since the 1850s. While Sir Richard Burton and David Livingston had become famous searching for the source of the Nile River in Ethiopia, other lesser-known explorers had died in obscurity looking for the terminus of the Awash River. More than one European explorer on camelback or on foot was ambushed and killed by fierce Afar and Issa tribesmen before the mystery was solved of where the Awash River came to an end—it flows northeast into the remote Lake Abhé.

By the mid-1960s, the explorers trekking through the Awash were geologists in Land Rovers and even an occasional helicopter. They were less interested in the course of rivers aboveground than in the movement of great blocks of the earth's crust belowground. The science of plate tectonics was emerging, and had given geologists a new way of looking at the topography of the earth. The vast Afar depression is a magnet for geologists, because it is one of only three places in the world where three giant plates of the earth's crust intersect, forming a giant Y. The two upper arms of the Y are filled with water—the Red Sea and Gulf of Aden. The arms meet in the Afar depression, a sunken triangle of land between the high Ethiopian plateau to the west and Hararghe plateaus to the southeast. The bottom leg of the Y is the Great Rift Valley.

Taieb headed to the southwestern part of the Afar in 1965, to the Awash valley, to find evidence of ancient lakes, not fossils. He was interested in the way water and the movements of the giant tectonic plates had helped sculpt the rippled terrain of the Awash over the past 2 million years, particularly at times when the climate was wet and the valley was covered with hundreds of lakes, like a sponge filled with water. Mapping the Awash was like surveying an area the size of the nation of Greece—the Awash valley extended for about fifty thousand square miles. But Taieb, who was born and reared in Tunisia, was well prepared for exploring the scorched earth of the Afar. He was familiar with the desert and had worked in the Sahara twice before, searching for petroleum. For him, the desert was "magic." When money was tight or roads came to an end, he would explore on foot with a local Afar guide, using donkeys to carry his supplies and sleeping under a mosquito net at night or staying with villagers.

By December 1969, on his third field season in Ethiopia, Taieb was surveying the northern end of the Awash valley with an Afar guide when he discovered the valley of Hadar. There were no roads where Taieb was going, so he drove two hundred kilometers along old tracks that had been made by Italian soldiers in the six bloody years in the 1930s when Mussolini's forces occupied Ethiopia. After the tracks disappeared, he followed tributaries of the Ledi and Awash Rivers, driving above them along the basalt-capped plateaus that separated the valleys. On the day he discovered Hadar, he had driven across one of these gravel-strewn plateaus when he came to an abrupt stop at the edge of a cliff. He looked at the valley below. What he saw was unforgettable—and would make his colleagues swear out loud in wonder a few years later when Taieb would take them to the edge of the rim and show them the valley below.

It was a geologist's nirvana. As Taieb stood on the rim overlooking Hadar for the first time, he could see the Awash River, fringed with trees, meandering through the beige-and-gray valley like a thick ribbon of green. Beyond it were badlands in every direction—as if the earth had been squeezed like clay into a landscape of crumpled hills and deep ravines that rose up steadily toward another plateau ten kilometers away. The arid terrain was chiseled here and there with sandstone pinnacles and deep ravines. What thrilled him most were the sloping walls of the ravines, which were like road cuts that extended for kilometer after kilometer, exposing layers of ancient red, yellow, and white dirt and rock. Taieb knew that those multicolored layers were made up of ancient sediments that had been laid down over millions of years and exposed by rain, wind, and the movement of the ground. If he could map those layers, tracing each one's path horizontally across the landscape across gaps, he could "read" the history of how the valley floor was laid down, layer by layer. Jon Kalb, an American geologist who would travel with Taieb to Hadar two years later, would describe the scene as "an enormous, flat-lying encyclopedia of natural history with part of one page exposed on this hill, another in that ravine, another on the crest of a ridge. The formidable task ahead of us was to put together the pieces and see how much of any one page we could read."

It was late in the afternoon when Taieb reached the rim and looked down upon Hadar for the first time. He knew that the sun sets fast near the equator, because its path is perpendicular to the horizon. But he could not resist exploring this compelling landscape right away. He was overwhelmed by what he found. Elephant bones and tusks were sticking out of the sandstone, along with the fossils of rhinoceroses and pigs—and those were just the fossils Taieb recognized (he was not trained as a paleontologist). He took a photo, collected an elephant tooth and a few other fossils, and drew a rough map of the sediments, labeling the layers of clay, basalt, and sandstone. "It was too rich," recalls Taieb. "I was alone and it was impossible for me to choose the fossils."

When he returned to Paris, he took the elephant tooth to the senior French paleontologist Camille Arambourg and his protégé, the paleoanthropologist Yves Coppens, who had inherited from Arambourg the leadership of the French team on the Omo Expedition in southern Ethiopia. Coppens took a look at the elephant tooth from Hadar and recognized it as the same species of elephant that had been found in the Omo sediments, which were conclusively dated at 3 million years. Elephant teeth are particularly useful for getting a quick fix on the age of a site, because they are large and, therefore, plentiful and well preserved. They also have features that change bit by bit over time, so paleontologists can sort them into distinct lineages of extinct elephants whose ages and preferred habitats are well known in Africa for the past 6 million years. This use of index animals such as elephants and pigs to date sites is called faunal dating or biochronology, and it is often used when there are no volcanic minerals to date the sediments where fossils are found— or along with radiometric dating as an independent test of the dates.

At 3 million years, Coppens noted, the elephant was older than the earliest known fossils of hominids coming from Africa (if he overlooked Louis Leakey's claim that the Miocene ape *Kenyapithecus* was a hominid). Coppens was intrigued, and asked Taieb to bring back more fossils from Hadar. The next season, Taieb pushed farther south into new terrain that was much harder to reach than Hadar. On that trip, he spent two weeks exploring the area on the western shore of the Awash River and passed by Aramis—areas that would become the epicenter of hominid fossil discoveries more than two decades later, in the 1990s, when Tim White and his Ethiopian colleagues worked there. But in 1970, it was inaccessible by Land Rover. Taieb borrowed a boat and cut through mosquito-infested swamps to cross to the western side of the Awash, where he spent two days exploring the sediments. He found fossils but left them in one spot, planning to return to retrieve them. Little did he know that he would soon be so busy at Hadar that he would not get back to the western side of the Middle Awash for many years—but he would tell his colleagues that he knew about a mysterious area that was likely to have fossils that were even older than those at Hadar.

Word of what Taieb was finding in the Awash quickly spread at the Seventh Pan-African Congress of Prehistory and Quaternary Studies, held in December 1971 in Addis Ababa, the capital of Ethiopia. It drew leading paleoanthropologists, including Louis Leakey, who at sixty-eight was in the last year of his life, and his wife, Mary Leakey. By that time, Taieb had surveyed parts of the Awash with Jon Kalb, who was also interested in plate tectonics and was living in Addis Ababa. The two showed the Leakeys the fossils of animals they had collected. Louis was characteristically encouraging, telling them the fossils were older than those at Olduvai and even expressing interest in going to the Awash himself. That would not happen—Louis died the following October, collapsing from a massive heart attack in a friend's apartment in London. The Leakeys did give Taieb and Kalb letters of support, which they used to raise funds for a preliminary expedition and to obtain permits. They also invited along another young American, Donald Johanson. Johanson was a twenty-eight-year-old paleoanthropologist who had studied with Clark Howell at the University of Chicago and had worked in the Omo for two seasons. He had met Taieb in Paris earlier that year, and though he had not finished his dissertation and had no job and only six hundred dollars in savings, he signed on.

Taieb finally had a team, funds, and permits for a two-month survey of the Awash. On April 22, 1972, he set off for the Awash valley with Kalb, Johanson, and Coppens, as well as several other researchers and a representative of the governmental organization that regulated archaeological research. On the day they got to Hadar, the team was tired from hours of driving over unpromising, featureless terrain. Late in the afternoon, Taieb drove across the plateau and stopped, without warning, on the rim overlooking Hadar. Johanson looked out his window, and was stunned. "It really took my breath away," Johanson recalls. "I thought, My Lord, this is enormous." Clearly, there was enough work to keep them busy for decades.

The first real field season at Hadar got off to a slow start. It was the fall of 1973, and half the field season had gone by without the discovery of a hominid. Don Johanson had all but promised the National Science Foundation in his grant application that he would find hominids at Hadar, and now he had spent almost half of his grant money of $43,000 on a Land Rover and supplies for the French-American team. (As a geologist, Taieb got even less funding.)

The money was supposed to last for two years. He was young, ambitious, and impatient for a hominid. Even though many paleoanthropologists have had illustrious careers without finding hominids and most teams spend years at a fossil site before seeing evidence of human ancestors, Johanson had yet to prove himself. He doubted that he could renew his grant to work at Hadar if the team found only fossils of pigs and hippos. Johanson was preoccupied with these worries on the afternoon of October 30 as he was surveying an airless little gully with Tom Gray, a graduate student in archaeology at Case Western Reserve University in Cleveland, where Johanson had just been hired as an instructor in anthropology. In a scene that Johanson described vividly in his book *Lucy: The Beginnings of Humankind,* Johanson idly kicked at what he thought was yet another bit of hippo bone sticking out of the sand. It came loose, and he realized that it was probably the tibia—the shinbone—of a small primate. Johanson picked it up, and as he was recording the spot where he had found it in his field notebook, he noticed another piece of bone a few yards away. This time, it was the lower end of a thighbone. Next to it was a bit of rounded condyle bone from the knee joint. He put the pieces together; they fit perfectly. At first, he thought he had found a monkey's knee.

Then he noticed that the thighbone and shinbone joined most naturally at an angle. He knew that a monkey's thigh and shinbones fit together one on top of the other, in a straight line. "Almost against my will I began to picture in my mind the skeleton of a human being, and recall the outward slant from knee to thigh that was peculiar to upright walkers," Johanson wrote. "It dawned on me that this was a hominid fossil." Other than isolated teeth, the knee bones were Johanson's first discovery of hominid bones. But the next morning, he began to second-guess his own judgment. He desperately needed a human thighbone for comparison. In a bizarre replay of Harvard paleontologist Bryan Patterson's raid of a Turkana grave, Johanson would also find a local tribe's burial mound and take a human thighbone. When he compared the fossil femur and the femur from a dead Afar tribesman, they were virtually identical, except that the fossil thighbone was smaller. He announced his discovery of the fossil at a press conference in Addis Ababa.

As soon as Johanson returned to Cleveland, he wanted confirmation that the knee was indeed from a hominid. He took the leg bones to Owen Lovejoy, a biological anthropologist at Kent State University in Kent, Ohio. Lovejoy was an expert on the biomechanics of locomotion. He had studied the legs of australopithecines, including a partial skeleton found in South Africa in 1947 that provided convincing evidence for upright walking in the Taung baby's species.

Lovejoy remembers the day that Johanson arrived on the doorstep of his house with a brown shoe box that said simply "Primate" on the side. Lovejoy, who has analyzed many famous legs from the fossil record, has a special fondness for the little knee that Johanson brought him that day more than thirty years ago—and he still has a framed watercolor drawing of the knee on display in the anteroom to his office. Johanson sat in Lovejoy's living room and casually opened the shoe box with four knee and leg fossils nestled in yellow foam. Lovejoy examined the bones, running his finger over the joint where the thighbone meshed with the lower leg bone. It was just like a miniature human knee. It was small, but it was not a child's knee, because the spongy bone found in parts of the knee joint in children had ossified, as in adults. This adult had been three feet tall or so, but it had walked upright. "The minute you find a distal femur (the lower end of the thighbone), you can tell a lot about how it walked," says Lovejoy.

The angle that Johanson had recognized was one way to tell that the bones had belonged to an upright walker. But the knee is the most complex joint in the body. It has many features that have changed as our ancestors adapted to upright walking, setting in bone telltale signs of the way the joint moved and the forces that acted on it. Humans can extend their legs fully, for example, unlike the African apes, which walk upright with a bent knee.

Lovejoy verified that the Hadar hominid had indeed walked like a human. Then he asked Johanson how old the fossils were. When Johanson told Lovejoy that a preliminary potassium-argon date suggested that the knee bones were about 3 million years old, Lovejoy laughed out loud. He knew that these leg bones would kick up a storm among paleoanthropologists when they realized that a tiny hominid the size of a chimpanzee had been walking around upright like a modern human 3 million years ago. This knee joint was confirmation for Lovejoy that australopithecines had walked the modern walk. The case was far from settled, however, and would be a subject of intense debate for the next thirty years.

Now that Johanson had an expert's opinion that his fossil walked upright and therefore really was a hominid, he could focus on his next problem. What kind of hominid walked on these leg bones? Was it a member of the genus *Homo*, which included humans and their direct ancestors? Or was it the knee of an australopithecine such as those found in South Africa, Kenya, and Tanzania, which some researchers thought were not direct ancestors of modern humans?

Before the American and French team could answer those questions, Johanson had a second big find—the partial skeleton known as Lucy. It was a year later, on November 24, 1974, and he was prospecting for fossils again with Tom Gray at Hadar. They were preparing to head back to camp at midday when Johanson decided to check out one last little gully. The first part of Lucy that he spied was a bit of arm bone lying on the ground, partway up a barren slope that had been the shoreline of an ancient lake. Next to it was the back of a small skull, and a few feet away was part of a thighbone. With increasing excitement, Johanson and Gray found vertebrae, part of a pelvis, and ribs. It didn't take long for them to realize that they were finding parts of a *skeleton*, which was unheard of at that time. No one had ever found a skeleton of a primitive individual before—and they knew from the preliminary potassium-argon date of the strata where she was found that she had lived about 3 million years ago, although the samples of basaltic rock from a lava flow were badly weathered and yielded dates that were not as reliable as the team's geologists wanted. (Eventually, new radiometric dates on better samples of potassium-rich volcanic rock would pinpoint Lucy's age to 3.18 million years.) Gray drove into camp blasting the horn on their Land Rover. Later that afternoon, the entire team fanned out over the gully to collect Lucy's bones—a task that would take three weeks and yield several hundred pieces of bone. And this time, the entire camp celebrated the discovery, drinking beer and listening to a tape of the Beatles' song "Lucy in the Sky with Diamonds" blaring over and over all night long. Sometime during that night of celebration, the ancient hominid that was taking shape on a camp table was baptized Lucy.

In the end, the team would find about 20 percent to 40 percent of

Lucy's skeleton. (The figure depends on what bones are counted—although Johanson's team found 20 percent of the 206 bones in the human body, he would say they found 40 percent of the skeleton, because he did not count the fragile hand and foot bones, which are seldom found.) Regardless, there was enough of her skeleton to make it clear that Lucy was something different. She was a mix of old and new, primitive apelike traits and modern manlike traits. She was neither ape nor human.

Her petite form also captured the public's imagination. Lucy would quickly emerge as an icon, perhaps because it was easier to imagine a female with a name like Lucy living on the shores of an ancient lake than the fragmentary fossils or skulls that had been discovered before her. Johanson and Coppens were suddenly in great demand on the lecture circuit in the United States and Europe. In English and in French, on television documentaries and in *National Geographic,* they painted a vivid portrait of a small female who was alive at the dawn of humanity. Johanson soon became a spokesman for paleoanthropology in the way that Carl Sagan was an important popularizer for astronomy.

Instead of describing billions and billions of stars, Johanson faced the task of describing hundreds and hundreds of fossil fragments back in his lab at the Cleveland Museum of Natural History, where he had been appointed curator of physical anthropology. And the fossils would keep coming—first in the form of four jawbones found the same season as Lucy; later, in a group of at least thirteen individuals who apparently died together in a catastrophic event, such as a flood. They would become known as the First Family. Studying the hominids from Hadar was becoming a population science, rather than the pursuit of an individual or two. But the question before Johanson and Coppens was, What kind of population did these individuals belong in? Were they apelike australopithecines or early members of *Homo,* or both?

CHAPTER SIX

DEFINING HUMANS

It may seem ridiculous for science to have been talking about humans and prehumans and protohumans for more than a century without ever nailing down what a human was. Ridiculous or not, that was the situation.

DON JOHANSON

Tim White thought the fossils from Hadar looked familiar the first time he saw them. In early January 1976, Johanson stopped in Nairobi with the fossils of the First Family on his way back to Cleveland. While he was there, he showed them to Richard Leakey and his mother, Mary Leakey, as well as White and other researchers who were working with them. By that time, Johanson and Richard Leakey were friendly. The two were about the same age (Johanson was thirty-two and Leakey was thirty-one), but Richard Leakey had the authority of early success—Johanson had admiringly called him the supernova of paleoanthropology. Johanson had sought his advice on fossil sites and on his fossils. He had even bet Leakey a bottle of wine that he would find older fossils at Hadar in 1973 than Leakey was finding at Koobi Fora. Johanson won that wager with his discovery of the knee bone, which was one million years older than a spectacular skull of *Homo habilis* that Leakey had found the year before at Koobi Fora.

At the meeting at the National Museums in Nairobi, Johanson spread a multitude of fossils out on a table as Richard and Mary and White crowded around. Richard Leakey had been finding fossils at Koobi Fora that looked like the earliest members of the true human line—including the skull of *Homo habilis,* which lived almost 2 million years ago. He and Mary also were finding more primitive fossils of hominids with smaller brains and larger teeth that they had taken to calling australopithecines. Almost all researchers split the fossils into two distinct branches in the human family tree: *Homo* and *Australopithecus.* Many researchers thought that one species of *Australopithecus*— *Australopithecus africanus,* which included the Taung baby—had given rise to the *Homo* line sometime in the past 2 million to 3 million years. But Richard and Mary Leakey, like Louis before them, thought the two branches had split well before 5 million years, and that australopithecines were not ancestral to the human line. The human line had much more ancient origins and bypassed the australopithecines completely, in their view.

Therefore, when the Leakeys saw the fossils from Hadar, they suggested that they belonged to two species of hominids. The more modern-looking members of the First Family and other jaws from Hadar belonged in *Homo.* But the primitive set of Lucy's jaw and teeth put her in *Australopithecus.* Johanson was at first inclined to agree with the Leakeys, who had years of experience and an air of infallibility. But White, who had been listening quietly to their discussion, made a surprising observation to Johanson at the end of the meeting. He suggested that the fossils from Hadar might be the same species as fossils that Mary Leakey's team had uncovered in Tanzania, and that they all represented just one species.

White knew the fossils from Tanzania well, because he was working at that time on Mary Leakey's fossils from a site known as Laetoli, which was about fifty kilometers south of Olduvai on the edge of the Serengeti plains. In 1975, shortly after Johanson found Lucy, Mary Leakey's Hominid Gang found teeth and a jawbone that were about 3.5 million to 3.7 million years old, which made them the oldest undisputed hominids. White had been describing the teeth and jaws of those fossils, and he compared them with the fossils from Hadar at the meeting in Nairobi. They were remarkably alike even though they came from sites that were a thousand miles apart. Johanson and White agreed to keep in touch. The next year, they agreed to work together.

Johanson wrote with self-deprecating candor in his book *Lucy: The Beginnings of Humankind* that White was initially wary of him at the meeting in Nairobi. Johanson felt he was generously sharing unpublished fossils he had found with colleagues because he wanted to learn from their comments. Debate sharpened his reasoning. But White had watched Johanson from the sidelines with characteristic skepticism. Johanson seemed to White like a "smooth, young hotshot." He had come out of nowhere in two short years since he had found his first hominid fossil, and he appeared to be moving too high, too fast. White was waiting for Johanson to slip up and say something naive. Photos of this meeting show Johanson with sideburns and a denim shirt, with a fossil cradled in his hand. White is sitting across from him, making a point about the fossils, while Richard Leakey, wearing a tie, appears to be seeking White's opinion. With sun-streaked blond hair that needed a trim and gold-rimmed aviator glasses, White is the only one dressed in a lab coat. At twenty-six years old, he was already a no-nonsense, data-driven scientist impatient to cut through the talk and get to the real work of making inferences from the fossils themselves.

But though they had different styles, Johanson and White shared one important trait: they both were self-made young Americans who took little for granted. Both were young, ambitious, and had worked hard to get to the fossil fields of Africa. Johanson had been born in Chicago, the only child of Swedish immigrants. His father was a barber who died when Johanson was only two years old. A neighbor who taught anthropology introduced Johanson to his books about cultural anthropology, opening up exotic cultures to the boy. When he was sixteen, he read Louis Leakey's account in *National Geographic* about Mary Leakey's discovery of Zinj at Olduvai in 1959, and around then he decided to become an anthropologist. His neighbor, however, advised him to consider studying chemistry instead, since he would surely starve as an anthropologist. Johanson followed that advice and initially majored in chemistry at the University of Illinois. But he switched to anthropology and began looking for mentors who could help him get to Africa. He read about the research of Clark Howell, who was then a young professor at the

University of Chicago, and decided to telephone him. Eventually, he transferred to the University of Chicago, where he earned his Ph.D. in anthropology. It was Howell who first made it possible for Johanson to go to Africa,
where he worked for two field seasons in the Omo valley.

White also was advised to forget a career in anthropology. He was told
it was impractical for a kid coming from the remote mountain community in
southern California where he grew up. In an essay in the book *Curious Minds:
How a Child Becomes a Scientist,* White wrote that he was the "luckiest kid in
the world," growing up on the edge of a national forest, where he learned
about wildlife and natural history firsthand. His father was initially a laborer
who worked for the county roads department; he knew the back roads well
and took his family exploring the mountains and desert on weekends. As his
father worked his way up, first to heavy-equipment operator and later to high-
way superintendent, they moved out of the small mountain cabin into bigger
houses, but they stayed in the undeveloped mountains near Lake Arrowhead.

White spent a lot of time alone, reading books—and some of his fa-
vorites were a Time-Life series on natural history, including one called *Early
Man,* written in 1965 by Clark Howell. White was nine years old when he
read the same article that inspired Johanson at sixteen and a whole generation
of future anthropologists—the *National Geographic* story on the Leakeys' discovery of Zinj in 1959. By the time he was in high school, he was trying his
own hand at archaeology, excavating Indian artifacts at one prehistoric site
with a friend and discovering, mapping, and cataloging others.

He was most interested in baseball and reptiles, and initially he wanted
to be a dinosaur paleontologist. But in his mountain community, he might as
well have said he wanted to be president one day or go to the moon. The
school counselor told him to be practical, and repeatedly reminded students
that no one from their mountain had ever graduated from the University of
California. So when White was accepted at the University of California at
Riverside, he was prepared to fail. Once there, he majored in biology and was
initially content with a C average. Then, inspired by a class with Professor
Wilbur Mayhew, a renowned expert in the field biology of southern California, he decided to add a second major in anthropology. This was despite the

fact that he walked out of his first course in introductory archaeology because the graduate student teaching the course relied on a textbook description of how to classify discoveries at an archaeological site, and was not receptive to White's protests that real sites where he had worked weren't patterned the way the book suggested. White graduated and went on to earn a Ph.D. at the University of Michigan, where he studied with Milford Wolpoff, a paleoanthropologist who, like White, is unafraid to take a contrary view or to say what's on his mind. Wolpoff also influenced White to consider differences in fossils' anatomy as variation within one species, rather than splitting them into different species, as Louis Leakey tended to do.

White had badgered Wolpoff to ask the Leakeys at a conference he was attending in New York in 1973 if they had room for him to work with them at Koobi Fora, so he could get some experience in the field. Richard Leakey agreed, and the next year White landed on the dirt airstrip at Koobi Fora. He was totally unprepared for how alien he would feel. "When I stepped off the plane at Koobi Fora and the place was flatter than I ever imagined, and the wind, the hot wind, hit me, I was surprised," White said. "It wasn't like what I had imagined on the pages of *National Geographic.*" He would still recall, thirty years later, feeling like an outsider: he traveled straight from the liberal campuses of California and Michigan right to the Leakeys' camp, which was like stepping back in time to a place where many of the old habits of colonialism still persisted. The renowned Hominid Gang of fossil hunters, mostly from the Kamba tribe, still ate and slept in different parts of camp from the Leakeys and the foreign researchers, and black Kenyans cooked the food and washed the clothes. White would apprentice himself with the legendary fossil hunter Kamoya Kimeu and others who taught him about searching for fossils. At night, he would often talk with them by the campfire. "They were asking all the right questions and had the right aptitude and interest to become professional scientists," recalls White, who felt uneasy with the inequities and thought that some of the Kenyans should be getting scholarships to study paleoanthropology. But though he felt like an outsider and was critical of many aspects of the way the camps at Koobi Fora and Olduvai were run, he was also the beneficiary of Richard and Mary Leakey's interest. Both noticed that he was bright and thorough as he worked on fossils. Mary gave him the opportunity to study the earliest fossils at Laetoli, where she was working.

White had finished his dissertation for his Ph.D. at Michigan when he began working with Johanson in Cleveland in the summer of 1977, almost two years after the Nairobi meeting. Both were hardworking, and their time together would prove a productive partnership—one that Johanson would boldly compare to that of Watson and Crick, the famous Nobel Prize–winning pair who discovered the double-helix structure of DNA. As they systematically worked through fossils and tried to decide how to classify them, Johanson would throw out an idea and White would play the devil's advocate. They talked endlessly about where to put the new fossils in the human family tree. Johanson's initial bias was to put most of the fossils in the genus *Homo*. But White relentlessly skewered that view in his discussions with Johanson, pointing to primitive features in the teeth and jaws of many of the fossils—apelike traits such as a protruding snout, a palate that was low and flat, and a small brain. In a notable scene in one television documentary, White showed just how small the brains of the new fossils were. In his surprisingly deep voice, White said in an offhanded way: "I've got an orange in my lunch that probably fits in that fossil." He pulled out an orange and fit it neatly inside a cast of a skull. Four oranges, on average, fit in the braincase of a modern *Homo sapiens*.

Although the earliest members of *Homo* still have relatively small brains, they have flatter faces than an ape and upper canines that do not sharpen against their lower teeth. Their jaws lack a gap between the canines and upper incisors. By contrast, australopithecines are more primitive and apelike. They had brains the size of chimpanzees and snouts that protruded more than in early members of *Homo*, but less than an ape's. What set them apart from apes, however, were their unsharpened canines and their lower bodies—they walked upright and could stand as tall as five feet.

Unfortunately, there was no clear-cut definition of what made a fossil on the cusp of early *Homo* or *Australopithecus* recognizable as one or the other. Where upright walking had been the traditional hallmark of being a hominid, brain size had been the traditional marker of being a member of *Homo*—until 1964, when Louis Leakey, British anatomist John Napier, and South

African paleoanthropologist Phillip Tobias named a small-brained fossil, discovered in 1960 at Olduvai, *Homo habilis*.

Johanson and White finally decided to score the fossils according to the same set of traits that Le Gros Clark had drafted in the 1950s to differentiate between apes and humans, and see where they fell on the ape-human yardstick instead of the *Homo-Australopithecus* continuum. By the end of the summer of 1977, they had logged enough differences to convince them that the fossils from Hadar and Laetoli were one species—and something new, warranting its own species. They decided that the fossils were australopithecines, rather than in the genus *Homo*, because they had small brains and lacked derived characters found only in members of *Homo*.

They baptized the new species from Hadar and Laetoli *Australopithecus afarensis* (meaning "southern man-ape from the Afar") and decided it was the ancestor of all the hominids that came later—both the australopithecines and the early members of *Homo*. This meant it was ancestral to the human line, which included *Homo habilis, Homo erectus, Homo neanderthalensis,* and *Homo sapiens*. It was also the ancestor of three types of australopithecines that came later: *Australopithecus africanus* (the Taung baby species), the aptly named *Australopithecus robustus,* and *Australopithecus boisei* (Zinj, from Olduvai).

It was the first new species of hominid to be named since 1964. The pair sent manuscripts to the journal *Science* and to the journal of the Cleveland Museum of Natural History. While they were in press, Johanson mentioned the new species name for the first time in public at a Nobel symposium held in May 1978 by the Royal Swedish Academy of Sciences in Stockholm. As Johanson described Lucy and the Hadar and Laetoli fossils, Mary Leakey was sitting in the audience, visibly perturbed, with a red face, according to several accounts. Even though she had agreed to be coauthor of the papers describing the new species (along with Yves Coppens), she did not accept placing the fossils in *Australopithecus,* which she still thought was not ancestral to humans. She was angry that Johanson had publicly described fossils her team had discovered, even though she had published an initial description in *Nature*. What bothered her most was that he had labeled them all with the name *Australopithecus afarensis,* with its nod to the Afar but not Laetoli. By this time, Mary Leakey's team had found more fossils of the new species at Laetoli, and she thought they belonged in the genus *Homo*. And researchers, including Tim White, had uncovered an ethereal trail of footprints in volcanic ash that dated to 3.6 million years and were the spoor of this new species. Johanson and White did not mention the unpublished footprints, but its trail would erase any lingering doubt that these hominids had walked upright.

Mary Leakey's response to the new designation was to ask Johanson and White to remove her name from a paper in press that would introduce the new species, even though it meant reprinting the cover page. This marked the beginning of the end of Mary Leakey's collaboration with Tim White and Don Johanson. White returned to Laetoli later in 1978, to work on the footprint trail, but he and Mary quarreled over the naming of Lucy and her kind, and the disagreement eventually turned into an unbridgeable rift.

Johanson and White were welcome almost everywhere else. Their article describing Lucy and the other fossils was on the cover of *Science* in January 1979, and they were featured in a special session at the annual meeting of the American Association of Physical Anthropologists in San Francisco soon after, where a standing-room-only crowd packed the ballroom of the Hilton Hotel to hear their presentation. Documentary filmmakers recorded the session where Johanson, White, Lovejoy, Bill Kimbel, and other members of the team sat in a row at a long banquet table onstage. After Johanson made it clear that they thought Lucy's species, *Australopithecus afarensis,* was the common ancestor of all the hominids that came later, a long line of researchers approached a microphone to ask questions. The startling nature of what Johanson and White had proposed became vivid when the paleoanthropologist C. Loring Brace of the University of Michigan took the microphone. He had been White's teacher, and had been to Cleveland to see the fossils. But he still held the old view of a single line of descent, where *Australopithecus africanus* gave rise to *Homo erectus,* with one species folding into another until modern humans emerged. He protested that the evidence was insufficient to "cast *Australopithecus africanus* into the limbo of the sidelines."

The new evidence would be subject to intense scrutiny, but Lucy would emerge unscathed—and most researchers would eventually accept her as the earliest known member of the human family. A few notable skeptics, including her codiscoverer Yves Coppens, have argued that Lucy's shoulder and arm bones show that she still spent so much time in the trees that she could not have been the direct ancestor of the first members of the human family who habitually walked upright—Coppens sees her more as a closely related contemporary of our upright ancestor, although this view has been hotly debated for three decades. Lucy's importance went beyond her status as the earliest known hominid (other than the unidentified fossils that Patterson had found). She also provided compelling new evidence to support one of Darwin's hypotheses a century earlier—that Africa was the birthplace of humankind. With the appearance of Lucy and the knowledge that different populations of hominids had lived a thousand miles apart at Hadar and Laetoli between 3 million and 3.6 million years ago, it was difficult to dispute that humankind arose in Africa.

While Lucy proved Darwin right on the Africa prediction, she proved him wrong on another proposal—that upright walking had evolved in concert with a big brain and the development of stone tools. The South African australopithecines had already challenged the notion that upright walking arose with a big brain, since the brain didn't begin its major expansion until much later—2 million years ago or so in the genus *Homo*. Darwin's idea had been retooled to drop the big brain and replace it with toolmaking as the important stimulus for upright walking. Man the Toolmaker needed his hands free, and this led to upright walking.

Then along came Lucy and other members of her species, walking upright with tiny brains—and well before stone tool kits show up in the fossil record, at about 2.5 million years. The thoroughly upright gait of Lucy suggested a host of new questions: How long did it take for Lucy's ancestors to develop a modern gait? Did upright walking appear rapidly in one population? Or did it take many, many generations to remodel the anatomy of hominids before they walked like a modern human?

Johanson was among those who recognized that it was time to consider new hypotheses about why human ancestors got up on their hind legs and started walking in the first place. "We now have to turn around and ask ourselves a new question," he said. "What prompted our ancestors to walk upright?" Owen Lovejoy would soon step in with some provocative ideas of his own—namely, that males needed their hands free to carry food to females, who would prefer to mate with males that provisioned them and their offspring with food. The better-fed females could more frequently have babies, and those offspring that were provisioned with more food would have a better chance of survival. Upright walking would provide a reproductive advantage and, therefore, be reinforced, or selected for evolutionarily. While this is one of the best-known and most comprehensive hypotheses for the origins of bipedalism, other researchers have also come up with scenarios where upright walking was advantageous for collecting food or carrying babies that could no longer grasp their mothers with their toes, as chimpanzees can with their opposable toes.

Lucy's timing was also superb. She arrived on the scene in the fossil record just as Pilbeam was beginning to realize that *Ramapithecus* was not a hominid. With her primitive looks at 3.18 million years, in the early 1980s most paleoanthropologists would finally reset their clocks to the time kept by the molecular clock, which dated the split of humans and apes to about 5 million years. One researcher who was particularly delighted by the arrival of Lucy was Berkeley biochemist Vince Sarich, who had been one of the first to recognize that dating the timing of the split of humans and apes was an interdisciplinary problem whose solution would require several lines of evidence from fossils and genetics. "I think that Lucy was the most important discovery, because Lucy looked awfully apeish in a number of her features, and yet she was around three million years old," Sarich recalls. "People said then, if you have something this primitive at three million, how much farther back do you have to go to get a common ancestor of chimpanzees and humans?"

CHAPTER SEVEN

BANISHMENT

Eating the bitter bread of banishment.

WILLIAM SHAKESPEARE

Lucy and her species set a new benchmark for paleoanthropologists seeking the earliest hominids. Once Donald Johanson and Tim White identified Lucy's kind as the earliest members of the human family, paleoanthropologists felt compelled to find her ancestors. Johanson described the time right before Lucy's species was alive—between 4 million and 7 million years ago—as a black hole in the fossil record. It clearly was exerting its pull on him: he was eager to return to Ethiopia to explore Hadar further. He was also intent on surveying the tantalizing fossil beds in the Middle Awash where Maurice Taieb had come across fossils he thought were at least 4 million years old.

This wasn't the only interesting gap in the fossil record of the human past—the period right after Lucy was alive, from 3 million to 2 million years ago, also was a mystery, and there were other gaping holes as well. Lucy's primitiveness, however, was alluring, and suggested to many paleoanthropologists that they were closing in on the right time and, perhaps, one of the places where the earliest members of the human family lived. But now that they had identified a promising time and place to search for Lucy's ancestors, they could not get there.

On Johanson's last trip to Ethiopia in 1977, a coup had taken place. In his book *Lucy: The Beginnings of Humankind,* he tells the dramatic story of checking out fossils at the Ministry of Culture in Addis Ababa from a friendly young man who was the permanent secretary. Releasing the fossils to Johanson would turn out to be one of the last acts of the secretary's life—he was shot that night while driving home from the ministry. A friend who was an attaché at the French embassy advised Johanson to pack his bags and spend that night in Addis Ababa at the embassy, where he would be safe. While he was there, Johanson learned from the ambassador that a new leader named Mengistu Haile Mariam had emerged to run the Derg, which was the mysterious council of military leaders that governed Ethiopia from 1974 to 1991. The deposed emperor Haile Selassie had died while imprisoned in his palace, and the general who had replaced him in an earlier coup had been shot and killed. Parts of the country collapsed into anarchy. On Mengistu's watch, Somalia invaded and occupied parts of Ethiopia, and civil war was waged in Eritrea, then still part of Ethiopia. The United States withdrew its support of the Ethiopian government because of its Marxist policies, backing Somalia instead. It became a dangerous time to be an American working in Ethiopia; American oil company employees were kidnapped and shot. Johanson canceled plans to return to the Middle Awash to hunt for fossils of human ancestors.

Johanson would not get back to Ethiopia until January 1980, when ministry officials began encouraging foreign scientists to work in Ethiopia again. When he returned as part of a small international delegation, he spent a little more than two weeks in the Middle Awash with Taieb and other geologists. They got tantalizingly close to the ancient fossil beds that Taieb had described. But a broken-down Land Rover prevented Taieb from taking Johanson to that remote spot.

The group returned to Addis Ababa, where Johanson and Taieb would hand over the entire Hadar hominid fossil collection—more than 350 bones. Johanson had kept Lucy locked in a wall cabinet in his office in Cleveland for five years. He knew, however, that it was time to return these fossils to the nation where they belonged and were considered precious antiquities. He would compare handing over her bones to giving away his own child to an adoption agency. But he was also focused on the future, and he wrote optimistically at the end of *Lucy: The Beginnings of Humankind* that he and White would return soon to Hadar and the Middle Awash to survey for hominids from these two gaps in the fossil record:

> What we find in them could well blow the roof off everything, because science had not known, and does not know today, just how or when the all-important transition from ape to hominid took place. This is the biggest remaining challenge in paleoanthropology. The gap between apes and us has narrowed in recent years, but it has never been shut. Lucy brings us close. She teaches us the astonishing fact that bipedalism goes back about four million years. But in her we see it already complete, with no clues as to how long it may have taken place. The feeling grows that one more step into the past will see its disappearance into a quadruped—into an ape.

Johanson would not get to take that step. White would instead go to the Middle Awash and those ancient sediments in 1981 with Desmond Clark, an eminent Berkeley archaeologist who had been working for years on a younger archaeological site nearby in eastern Ethiopia, and in other parts of Africa.

In a remarkable turn of events, Clark had gotten the permit for the terrain that Taieb had first explored, including the area that promised to be older than Hadar. Taieb had initially held the permit to the Middle Awash, but he'd lost it in the mid-1970s to the American geologist Jon Kalb, who had been part of the original team of French and American researchers that went to Hadar. Kalb resigned from the team because he and Johanson did not get along, and then managed to wrest the permit for the Middle Awash away from Johanson and Taieb, who were left with Hadar and the area around it. Kalb headed to the Middle Awash for several seasons in the mid-1970s and was the first one to explore the fossil beds at Aramis in the Middle Awash. His team found fossils of animals that were 4 million to 5 million years old, along with a striking skull of an early human that was alive much more recently than Lucy—about 500,000 years ago. Called the Bodo skull, it shed light on the mysterious transition from earliest hominids to archaic *Homo sapiens*.

Kalb's time in the Middle Awash came to an abrupt end in August 1978—he was expelled from Ethiopia amid rumors that he was associated with the CIA. Kalb described the dramatic events that led to his expulsion in his memoir *Adventures in the Bone Trade*. He adamantly denies that he worked for the CIA, but the CIA, as a matter of course, will neither confirm nor deny the rumors. With only six days' notice, Kalb was forced to evacuate Ethiopia with his two young daughters. He and his wife had lived there for seven years and planned to stay indefinitely.

Almost a decade later, Kalb won an out-of-court settlement from the National Science Foundation, the main source of funding for Americans doing research in human evolution. NSF officials admitted that their program officers had gossiped about Kalb's rumored connection with the CIA in a session where his application for research money was being considered; the NSF ultimately denied him funding. Kalb later wrote that some of the researchers who had alerted the NSF about his possible CIA connection were his competitors for the permits and funding to work in the Middle Awash. As part of Kalb's settlement, the NSF issued a statement saying, "The rumor has no basis in fact. No documentation or any other type of evidence has ever been produced to support the CIA rumor." Even the office of Vice President George H. W. Bush got involved, seeking a review of Kalb's complaints about the NSF. In a report to Bush, the NSF's director of audit and oversight Jerome Fregeau observed in 1982: "The exceedingly cut-throat level of competition in Eastern African anthropology is a long-standing problem. NSF cannot be blamed for it, but it must be kept in mind when making decisions on proposals in this area."

But the NSF's heightened awareness of the internecine conflict between paleoanthropologists vying for the same funds and turf came too late for Kalb. When he was expelled in 1978, he was shut out of the fossil-rich areas of Ethiopia for two decades. He returned to Texas, where he is a researcher at the University of Texas in Austin.

Not surprisingly, others were poised to move in quickly to gain the permits to work in the parts of the Middle Awash that had been Kalb's territory. A senior collaborator of Kalb's had already invited Clark at Berkeley to lead the work in the Middle Awash in 1977 when Kalb was unable to get funding—before Kalb was expelled from Ethiopia. Clark submitted an application for funding to the National Science Foundation, which was granted. In January 1980, Clark, Johanson, and an NSF program officer met in Addis Ababa with an Ethiopian government official and two French researchers, including Taieb. Their plan was to form a working group to develop a research program in the Middle Awash, and they were jointly given the permit for the Middle Awash.

Shortly after, in yet another byzantine twist, Johanson was charged with trying to steal a fossil that he had allegedly put in his pocket. Taieb defended Johanson, telling the museum worker who made the charge to look in Johanson's pocket, where he found a pencil. To this day, both Taieb and Johanson say the charge was ridiculous and a setup—why would either one steal a fossil that they had already returned to the museum, especially when they could have taken any fossils they wanted much more easily in the field? Why would they sign them in, only to steal them later? In the meantime, Clark ended up with the permit for the whole area, because he was the only one of the researchers named on the permit left. Johanson would temporarily lose his permit to work at Hadar. Eventually, he regained the permit, but Taieb and Coppens would give up. "I got tired of the politics," Taieb recalls. "It was very difficult." He wanted to work on other projects in Kenya and Tanzania.

Clark went to the Middle Awash in the fall of 1981 with White, then a young faculty member at Berkeley. Taieb had told White about the sediments that were older than those at Hadar. The Berkeley team had a successful field season, and they held a press conference in Berkeley where they announced the discovery of new fossils. They had found a thighbone and skull fragments that they identified as australopithecines that, the team estimated, had lived 3.9 million years ago.

Clark and White returned to Addis Ababa in August 1982 for a second field season, this time with White as coleader. Johanson also finally got permission to return to Hadar. But just before the two teams were to leave for the field in mid-September from Addis Ababa, the Ministry of Culture told them that their field permits were suspended—indefinitely. The official reason was that Ethiopia needed to rewrite its policies regulating prehistory research. By that time, the fossils of hominids and artifacts that had been found in Ethiopia were scattered all over the world, under study or even forgotten in various laboratory and museum drawers in Europe and the United States. They were a unique and irreplaceable part of Ethiopia's heritage, and government officials recognized that they needed to bring those fossils back home, as Johanson had done with Lucy. They planned to write regulations in which fossils and other antiquities would be recognized as the property of Ethiopia and could not be removed without special permission, as was the case with fossils found in most other nations.

There was also another unofficial reason for the moratorium. Kalb's former students who were Ethiopians had complained to government officials that the American and French teams had excluded them after Kalb was expelled, and that those teams had not done enough to train Ethiopians to become paleoanthropologists. Kalb would later help four Ethiopian students apply to graduate programs in the United States. Clark, in fact, also had a long tradition of recruiting and training African students at Berkeley from the nations where he worked, including Zambia, Malawi, and Ethiopia. And Clark and his Berkeley colleague Clark Howell had secured NSF funds to build a new paleontology lab at the National Museum in Addis Ababa in 1982. In the end, however, the foreigners' infighting and territorial skirmishes backfired. The Ethiopian students' complaints resonated with officials who already feared that they were being taken advantage of by ambitious American and European scientists.

Johanson, Taieb, Clark, and White responded by enlisting the support of powerful allies, including university chancellors and foreign dignitaries. But the Ethiopian officials would not yield to international pressure. That same October, the Derg made it absolutely clear that there was to be no field season that year or anytime soon: it issued a complete suspension of all prehistory research in Ethiopia. The moratorium on fieldwork lasted until

1990—eight years. The American and French researchers could not search for more fossils; but they could take another Ethiopian resource back to the United States and Europe with them—Ethiopian students. The Berkeley group in particular intensified its efforts to recruit and train Ethiopians. Several would become key members of their scientific team and critical to its future success in Ethiopia.

In 1974, the same year that Don Johanson found Lucy, a British graduate student named Martin Pickford found a molar that was twice as old as Lucy's remains. Pickford was mapping the rugged terrain of the Tugen Hills in Kenya as part of his thesis for a Ph.D. in geology. The Tugen Hills are a rare spot in the Great Rift Valley where a giant block of land, called a tilt block, was pushed up to expose layers and layers of ancient sediments. The exposed layers are a time machine for researchers, opening a window into the past from 16 million years ago to modern times—a time span never seen in one place before in the fossil record of Africa. As a result, geologists and paleoanthropologists were drawn to the Tugen Hills for more than a decade before Pickford arrived. The renowned British geologist William Bishop had joined an ambitious project by British geologists to map the terrain in the Tugen Hills and the adjacent valley around Lake Baringo, a muddy brown lake known for its crocodiles, hippos, and hundreds of species of exotic birds. Bishop dispatched two students, including Pickford, to look for fossils in particular.

As Pickford traced the contours of ancient sediments across the red hills and dry gullies, he mapped fossils eroding out of the hillsides in more than two hundred places. He wasn't the first one to find fossils in the area. But one day in 1974, when he spotted a molar in sediments that were about 6 million years old, he recognized it was important. "I always said it was a hominid and it was twice as old as Lucy, but it didn't make a big impact," said Pickford.

The tooth appeared in *Nature*. It was a significant accomplishment for a thirty-one-year-old graduate student in geology, but the report written with the paleoanthropologist Peter Andrews from the British Natural History Museum didn't get much attention in the popular press. It was the oldest known fossil that might have belonged to a hominid, excluding *Ramapithecus,* which still was a contender for hominid status. But it was found the same year that Don Johanson, also age thirty-one, discovered Lucy. A single molar, no matter how old, was a paltry tidbit compared to a partial skeleton. And without any other bones, it was difficult to prove the molar's identity as a hominid's. Some thought it was a hominid tooth. Others thought it belonged to an early chimpanzee or an extinct ape. Regardless of its identity, the molar was interesting because of its promise—it was another scrap of evidence that the ancestors of humans or chimpanzees were indeed living in eastern Africa more than 4 million years ago.

It would be years before researchers would find more fossils of hominids in the Tugen Hills. Pickford focused more on the geology for his thesis, and collected thousands of animal fossils between 1975 and 1980. At that point, Pickford wrote a monograph on the animal fossils, but he would say later that he was trying to resist jumping into the "rat race" of finding early hominids.

By the time Pickford found the molar, he already had been a close observer of the quest for human ancestors for more than a decade. He had been a close friend of Richard Leakey's as a teenager and had spent a year working as a personal assistant to Louis Leakey, helping to arrange logistics for expeditions and catalog fossils of pigs, apes, and other animals from research sites. Like Richard and Louis Leakey, Pickford was adept at exploring the wild terrain of Kenya, where he had lived since he was three years old. He had spent most of his childhood on a farm near Kitale, not far from the Tugen Hills, where his father raised cattle and sheep. Pickford's backyard was the Great Rift Valley, where he'd unconsciously prepared to become a naturalist, collecting rocks and fossils, learning the names of different birds, plants, and animals, and mingling among the local Kalenjin people, whose children were his playmates and whose language he still speaks. When he was eight years old, he saw Mount Longonot, a classic volcano with a gray cone that dominates the view of the Great Rift Valley along the main route northwest of Nairobi. He wanted to know how it was formed, and he began to think about becoming a geologist.

As a teenager, he boarded at the Duke of York School in Nairobi, where Richard Leakey was a day student a year behind him. The two were friends, as were their brothers, and Pickford occasionally spent the night at Leakey's house on weekends. "We were alike. We were into animals, natural history, and got along with Africans," recalls Pickford. "We didn't dare admit that our friends out of school were [black] Africans or we would have gotten a hiding by our [white] classmates." After high school, Pickford joined the British army, which honed his sense of discipline and organizational skills. It also taught him that he could not count on other people, he would say years later. After a year working for Louis Leakey, Pickford decided that he wanted to study geology, and his interest in animal fossils followed naturally. He enrolled in Dalhousie University, in Nova Scotia, where he studied with the South African paleontologist Basil Cooke, known for his research on fossil pigs in the Omo valley of Ethiopia.

Pickford next moved to London to earn his Ph.D. with geologist William Bishop at the University of London's Bedford College. Before long, he met another British student, Andrew Hill, who was Bishop's first student to study the fossils and ancient environment of the Tugen Hills and who had already found some australopithecine remains near Lake Baringo. The two took an almost instant dislike to each other. Pickford blames their rough start partly on Bishop, who, he says, told Hill he was giving his grant to Pickford. Hill dismisses that explanation. He puts it simply: "He wasn't the kind of person I liked. He didn't like me either."

Their distaste for each other could have faded into a distant memory, except that it would be revived over and over again as their paths crossed many times during the next three decades. As graduate students, both took an interest in the Tugen Hills. Both would earn their Ph.D.s in 1975. Both would end up working in the National Museums of Kenya. Both would also work with Yale paleoanthropologist David Pilbeam in Pakistan, searching for fossils of *Ramapithecus* in the 1970s, and later in the same fossil beds in the Tugen Hills. And both would find fossils of hominids there—but not together.

Pickford was a maverick from the start—fiercely independent and clearly used to keeping his own counsel. He can be charming, regaling a group with stories from his adventures in far-flung places. He can also connect with the local people where he works in Africa, drawing on his lifelong familiarity with Africa and speaking Swahili and some Tugen, which is spoken by the people of the Tugen Hills. But he keeps his distance from many of his colleagues in human paleontology, and sometimes has the cynical affect of a researcher who has been outside the scientific establishment for many years. He is on the faculty at the Collège de France in Paris, but he says he is most content in the desert or in the remote country of Kenya, away from the intensity of cities and academic politics.

He clearly prides himself on his self-sufficiency and his ability to manage tough logistics. He is highly disciplined, up at dawn and impatient with those who do not work as hard as he does: "You have to be prepared to sweat, to build tracks, to walk everywhere. You have to be part explorer, part road crew, part camp boss." He knows the geology intimately in the sites where he works, and he has a proven track record for finding fossils in the field, often with his longtime partner, French paleontologist Brigitte Senut of the National Museum of Natural History in Paris. Together, they look like an unassuming middle-aged couple: Pickford's beard is turning gray, he wears his glasses on a chain, and his khaki pants are invariably wrinkled; Senut has cut her salt-and-pepper hair short, uses little makeup, and wears sensible shoes, even in Paris. But there is nothing stodgy about either of them—they clearly live for their work and thrive on adventure and pushing the limits. Where Pickford is pragmatic and states that he has had to develop the "hide of a rhino," Senut can be emotional, erupting easily into laughter or tears, as she recalls the good times—and bad—they have shared as a team. It has not been easy being a woman in the male-dominated world of paleoanthropology: She alternately expresses frustration and pride at the obstacles she has faced and overcome, both academic and logistical. She grew up in a suburb of Paris in a family with three girls; she was the "boy" of the family, the one who was interested in rocks and geology and who went to the University of Paris to train to be a teacher in Africa. While there, she heard a lecture by Yves Coppens on human evolution, and after obtaining a master's in geology, she eventually earned her Ph.D. in paleontology, studying with Coppens to become a specialist on arm bones, in particular. But for her, the "best school is the field."

After dinner in camp, Pickford and Senut segue from one story to another, describing how they survived a coup in Uganda by hiding for several days in a schoolhouse, and how they got shot at by cattle rustlers. They still laugh about the time Senut heard a lion roar nearby and realized that she had been left alone at a fossil site in Uganda, unarmed. The field-workers had quietly retreated to the vehicles. After Senut escaped the lion, whose roars got louder as she walked to her car, the field-workers explained that they had not worried about her because lions don't eat white women.

The self-determination that serves Pickford and Senut well in the field, however, has made Pickford, in particular, enemies in paleoanthropological circles—and Senut has suffered by association with Pickford, to whom she is deeply loyal. Pickford is blunt, says what is on his mind, and doesn't worry about what other people think about him. He trusts his own judgment, for better or worse. He has criticized colleagues for making mistakes in scientific publications; yet more than one colleague has said that it was *he* who was mistaken: a Dutch paleontologist wrote in a science journal that it was "a habit" of Pickford's to accuse him of errors he did not make (although Pickford responded that criticism of published work is part of scientific debate). Pickford has also burned bridges with more than one colleague in disputes over access to fossils and fossil sites in Kenya, Namibia, and Uganda. His friends say he can be oblivious to the impact of his actions on others, but that he has been ostracized by powerful paleoanthropologists because he challenges their authority and doesn't bow to their rules for working in eastern Africa. His enemies are less charitable, calling him a pariah they do not trust because he seems hell-bent on revenge against competitors who he says have done him wrong—namely, Richard Leakey, Hill, and David Pilbeam. Pickford, however, dismisses their criticism and says simply: "I am not the devil." He sees himself as the victim of what he calls "the Lobby," a powerful group of paleoanthropologists primarily in the United States who, he says, have tried—and sometimes succeeded—to block the funding of his fieldwork and the publication of manuscripts he has submitted to journals.

Hill, by contrast, is mild-mannered, with a wry sense of humor, and mixes well in academic and paleontological circles. He is tall, with black hair and a salt-and-pepper beard, and dresses with a bit of urban flair. He favors bright shirts and a leather jacket, which are upscale among paleoanthropologists, whose standard uniform is khakis or jeans. A professor of anthropology at Yale, Hill is known for his hospitality—he is a gourmet cook, plays the harpsichord, and lives in a historic farmhouse in the countryside of Connecticut with his longtime partner, Sally McBrearty, a professor of anthropology at the nearby University of Connecticut.

Like many of the paleoanthropologists of his generation, Hill was inspired to take up the study of hominid fossils when he heard about the Leakeys' discovery of Zinj in 1959. He was a boy in Mansfield, England, when he watched Mary Leakey, on television, describe finding Zinj. Searching for fossils in Africa seemed like a remote prospect, and he majored in geology instead, at the University of Reading. Bishop, who also would become Pickford's adviser, then came to Reading to give what promised to be a "boring talk on mud in the middle of England," according to Hill, who said he would have liked to skip the lecture. But one of his professors wanted to introduce him to Bishop. Hill's interest was captured when Bishop said he needed a geologist to look at fossil sites near Lake Baringo. The following June, in 1968, Hill was working in the Tugen Hills.

Hill quickly developed an interest in how the fossils of animals were buried in a site and preserved—a science known as taphonomy. For several years in the late 1970s, he lived in Nairobi, where he worked for the National Museums of Kenya. Hill eventually settled across the Atlantic, at Yale. Now its chairman of anthropology, he has a large light-filled lab, a steady stream of promising students to mentor, and dining privileges in a historic Ivy League eating club. From this comfortable position he pursues his varied research interests, notably how changes in the global climate might have altered the habitat in the Tugen Hills and other parts of the Great Rift Valley when the first hominids were emerging. He is steeped in the details of the history of paleo-anthropology, and was recently curator of an exhibit on the past century of discoveries about human evolution at Yale's Peabody Museum. Hill's team has found its share of fossils of ancient apes and a hominid in the Tugen Hills, but none have been spectacular discoveries. Hill did not include a photo of himself in the exhibit (though fossils his team has found are featured). Nor does he exude the intense drive or single-mindedness that marks Pickford or many of the discoverers featured in the exhibit.

Despite their obvious differences, Pickford and Hill would agree on one thing: that the mid- to late Miocene, from 5 million to 17 million years ago, was one of the most interesting times in the history of primate evolution. Both underwent an intellectual indoctrination of sorts when they worked in Pakistan in the 1970s with Pilbeam. By that time, Pilbeam had come to think that paleoanthropologists should approach their sites like geologists who drill cores in the ocean floor at different sites around the world to see what the cores show about the climate at the same time in different places. Pilbeam wanted to know what the world was like in Pakistan, Kenya, Turkey, Europe, and China when a few apes were standing up on their hind legs and becoming bipedal hominids. Perhaps clues in the environment, coming from the plants and animals that lived alongside them, might reveal why one group of primates became hominids while others continued to evolve as apes. "We need to reconstruct past species as though they were alive today, as though we were watching them in a Time Machine," he wrote. Pickford, Hill, and the French paleontologist Michel Brunet, who also worked in the Siwaliks with Pilbeam, shared Pilbeam's deep interest in the late Miocene as a critical time for finding the earliest hominids, particularly after *Ramapithecus* was bumped from the human family.

After the Siwaliks, Pickford and Hill would have preferred to go their separate ways, but Pickford ended up working at the National Museums of Kenya, where Richard Leakey was the director from 1968 to 1989. Leakey had hired Hill in 1975 to work as a researcher, but not Pickford. Pickford, who wanted to stay in Kenya, was finally hired in 1978 by the Kenyan historian Bethwell A. Ogot, who was the first director of The International

Louis Leakey Memorial Institute for African Prehistory (TILLMIAP), an independent research institute at the National Museums, where Hill also would work. But Ogot was forced to resign in 1980, during a tense transition in which David Pilbeam replaced Ogot as acting director. In his autobiography, Ogot blamed Leakey for his ouster and Pilbeam for helping to enact it. But Pilbeam said a panel of consultants outside of the museum conducted an inquiry and that they, not Pilbeam, recommended that the attorney general seek Ogot's resignation (Pilbeam, incidentally, had been invited by Ogot in 1978 to be a scientific advisor to TILLMIAP). At the time Ogot resigned, Richard Leakey was extremely sick, awaiting a kidney transplant in London.

At about the same time, in 1980, Pilbeam won a grant to establish the Baringo Paleontological Research Project, which included the Tugen Hills, and asked Hill to be the field director. Even though Pickford had found—in the Tugen Hills—the Lukeino molar and a fragment of an arm bone of a primate that was about 4.5 million years old, Pilbeam did not invite Pickford to be part of the Baringo basin project. Pickford was no longer a member of the Siwalik project in Pakistan, and the goal of the Baringo research was to compare the fossil sites in the Siwaliks directly with those of the same age in the Baringo basin. Also, Pilbeam added, he was well aware that Pickford had had "difficulties" with several members of the team in the Siwaliks, and Pilbeam wanted to avoid the same problems in Kenya. Instead, Pilbeam recommended Pickford for another expedition in Greece, but it never materialized.

It would become a sore point for Pickford, who thought it was an effort by Leakey and Pilbeam to divert him from the Tugen Hills, where he had worked often from 1972 to 1980. Pilbeam denies this and says he was genuinely trying to do Pickford a favor (he also wrote him letters of recommendation for grants and research positions at the time). Pickford became even more annoyed when, upon Pilbeam's return to the United States, Hill took over the management of the Baringo project in 1980 and became an administrator at TILLMIAP after Ogot's resignation.

Under Hill's management, a member of the Baringo project, Kiptalam Cheboi, found a jawbone of a hominid in the Tugen Hills that was about 4.5 million years of age. Cheboi, who grew up in the Tugen Hills herding goats, has an uncanny eye for fossils and was working with Hill and McBrearty on the day in 1984 when he picked up the fragment of a jaw. Hill felt confident it was a new type of hominid, but he was reluctant to give it a new species name, since it was only a bit of jawbone with two molars and some roots. But in a report in *Nature* in 1985, Hill wrote that the new fossil from Tabarin established the presence of the hominid family in Africa as much as 5 million years ago, which, he noted, was a million years older than Lucy's species, *A. afarensis*. The specimen resembled Lucy's species more than any other type of hominid, but it was obviously older. Hill remarked that, at that age, the Tabarin fossil was getting nearer to the divergence point between hominids and the African apes. Today, he recognizes it as a member of the genus *Ardipithecus*, the same genus as the skeleton Berkeley paleoanthropologist Tim White and his colleagues would discover a decade later in Ethiopia.

Hill was eager to push back in time and explore fossil beds that were a bit older—particularly in the Lukeino formation, which was 5.6 million to 6.2 million years old and where Pickford had found the hominid molar a decade earlier. No one else was working in sediments of that antiquity in Africa at that time, partly because only a half dozen sites with fossil beds of that age were known—and those in Ethiopia were then off-limits to foreigners because of the moratorium. The stage was set for the entrance of a new player in the saga of human evolution—one that was older than Lucy—and Hill knew it. Hill spent a season working in the Lukeino formation fossil beds, but his proposal to focus there further was shelved when the team opted instead to concentrate on much older sediments in the Ngorora formation. This was between 12 million and 8.5 million years old, and covered another gap in the fossil record of Africa. The team had found an important partial skeleton of an extinct ape called *Equatorius* that had lived 15 million years ago, which kept them focused on the older slice of time. Hill would later regret that he did not focus more on the Lukeino sediments.

Pickford also was eager to get back to the Lukeino formation in the Tugen Hills. He had made it known to Richard Leakey that he still had a proprietary interest in the area and wanted to get a permit to work there so he could complete a monograph on the geology of the Baringo basin. But

Leakey, who had the authority as museum director to recommend approval or rejection of all permits to search for fossils in Kenya, refused. Pickford had been sent—as a geologist—to make a detailed inventory of archaeological, paleontological, and monument sites in Kenya. But he was told by Leakey not to collect the fossils he found in the field—he had to leave them in place for paleoanthropologists to collect. Pickford felt excluded from paleontology research by Leakey and complained. In 1984, Leakey did not renew Pickford's contract. According to Leakey, Pickford was an expatriate and had been hired with the understanding that he was to train an African to take over the job. Pickford, who was the father of four children, was out of work.

The next year, in 1985, Pickford began collaborating with Senut, who as Coppens's student had studied Lucy's arm bones in Cleveland while Johanson and White were doing their initial analysis. She was particularly interested in the origins of upright walking, but she disagreed with Lovejoy's analysis that Lucy had walked in a modern manner. Her view was that Lucy's species was off the main line to humans—a minority view that Coppens and a handful of other researchers share. Her distinctive interpretation of the fossils puts her at odds with many colleagues—particularly Tim White, who, she said, did not treat her with sufficient respect after a guest lecture at Berkeley in 1982 when she discussed this view. "He showed me bones and tested me as if I was a student," she recalled. "I would never have done that to a colleague." (White recalled the incident differently: he said he wanted to show her that there was variation in one trait within modern humans, and suggested she look at the collection of casts of the bone in the lab at Berkeley. When Senut said she didn't need to see them, White told her that if she were his student, she would be required to study them.)

By the time she began working with Pickford, Senut also felt outside of the mainstream of paleoanthropology and resentful of what she considered the arrogant American dominance in the field. But she persevered with such hard work and passion for the fossils that her mentor, Coppens, compared her to Joan of Arc when he gave her a research award in Paris recently. Senut had good connections in Europe, where a half dozen colleagues sympathized with Pickford and offered him various positions lecturing in Germany, Italy, Spain,

Portugal, and France. Together, they got permission in 1985 to work in Uganda, where they began a project to revisit fossil sites surveyed by Pickford's adviser Bishop.

~~~~~

On July 2, 1985, Pickford and Senut were in Nairobi, en route to Uganda, when they decided to stop at the National Museums of Kenya. While they were there, they had an encounter with Hill and Leakey that would set in motion a sequence of events that for years would wreak havoc in all of their lives. Hill, Leakey, and Pickford disagree on the details, but Pickford says he was at the museum to study Bishop's notebooks from a fossil site called Napak in Uganda, where 19-million-year-old fossils of Miocene apes were found. Bishop's wife, Sheila Bishop, had donated the notebooks to the museum archives after his death in 1977, and Pickford asked Leakey if he could look at them. Pickford says he took them out of the museum so he could photocopy them after signing them out in the logbook. The next day, July 3, Leakey wrote Pickford a letter charging him with attempting to steal documents—Bishop's notebooks—from the museum's archives. The letter also banned him from the museum's research facilities. The ban effectively meant banishment from ever working in Kenya, because at that time the Kenyan government would not issue a research permit to work in Kenya without the National Museums' recommendation—and, hence, Richard Leakey's signature.

Pickford implicates Hill, who, he says, was spying on him while he was working on Bishop's notebooks. Hill, who was on the faculty at Yale at that time and also visiting the museum on his way to the field, says he was not spying on Pickford. But he says he did not trust Pickford with the notebooks, because when he was at the museum on the Saturday before their encounter, he had moved a cabinet away from a wall to search for an electric converter for his typewriter. Instead, he found a pile of Bishop's papers on the Ugandan fossil site stuck underneath a bench. He was surprised to find them there because he had helped negotiate their donation. He had been systematically going through offices to find an adapter plug, and he realized then that he was in Pickford's former office. He wondered if Pickford had hidden them there earlier so he could study them on his return. (Pickford says everything was removed from his office after he left, so he could not have hidden the papers.) Regardless, Hill didn't say anything about it to Leakey. Instead, he put them in his own office to see what would happen, because he knew Pickford and Senut were in town.

The following Tuesday, Hill walked into the Department of Sites and Monuments to get a book while Pickford was going through Bishop's archives. He nodded at Pickford, then returned to the vault where he was working with Richard Leakey on fossils. He casually mentioned that he had seen Pickford. Richard immediately stopped what he was doing and rushed out of the hominid vault, Hill recalled. Pickford had left, and the head of the department told him that he had signed out papers from the archive to photocopy. This did not make sense to Richard, he said later, because there was a photocopy machine in the library. Pickford says it was out of order that day.

Richard sent Pickford a letter charging him with theft, but he gave Pickford the opportunity to return the papers. According to Richard, Pickford refused, and Richard later reported the incident to the museum's board of trustees. Pickford insists he returned the papers that day. But all eleven trustees eventually voted to ban Pickford from working at the museum. At that point, Hill told Leakey about the pile of Bishop's papers he had found in Pickford's old office. "I wasn't spying on him, but it's true I was suspicious of him," Hill said.

The ban was devastating to Pickford. Over the next nine years, he would repeatedly write to Leakey, to the museum's trustees, and to government officials to clear his name. Richard Leakey told Senut that the ban did not apply to her, even though she worked closely with Pickford and they had shared permits. But she stood by Pickford's account of events, and would not work again in Kenya without Pickford. By then, Senut and Pickford were living and working together. She would contact Coppens, and in 1993 he would help Pickford get an appointment as a senior lecturer at the Collège de France in Paris. Pickford thought about suing the Leakeys, but his lawyers in France told him he would need a million dollars to proceed. "I was banned from working in Kenya. It pissed me off, and it made me more determined to return." The banishment of Pickford would reverberate in paleoanthropological circles for the next two decades.

The

Decade

of

Discovery

*New and significant prehuman fossils have been unearthed with
such unrelenting frequency in recent years that the fate of any
lecture notes can only be described with the watchword of a
fundamentally irrational economy—planned obsolescence. Each
year, when the topic comes up in my courses, I simply open my
old folder and dump the contents into the nearest circular file.
And here we go again.*

STEPHEN JAY GOULD

———————

CHAPTER EIGHT

THE LADY OF THE LAKE

The real voyage of discovery consists not in seeking new landscapes but in having new eyes.

MARCEL PROUST

Meave Leakey first saw Kanapoi from the air. She was flying over the parched badlands of northern Kenya in a single-engine Cessna with her husband, Richard Leakey, when she asked him, "Where's Kanapoi?"

Richard, who was at the controls, searched for a familiar landmark. He found two dry rivers whose paths were etched in the beige desert and lined with a thin fringe of green acacias. He traced the sand rivers to a spit of land where they met near a ridge with striking white layers of ash. "That's Kanapoi," he said. "There's nothing there."

Or so he thought. He had been there on a doomed scouting mission three years earlier, in 1981. With a small team of fossil hunters from the National Museums of Kenya, Richard and paleontologist Alan Walker, his longtime colleague, had set out to explore the hills on the west shore of Lake Turkana, an immense jade-green lake that cuts 180 miles down the center of the Great Rift Valley, from the Ethiopian border into northern Kenya.

By the time Richard set out to scout the western shore of the lake, he was already famous for his discoveries of fossils on the eastern shore, where he had returned every year since 1968. He had discovered the rich fossil beds at Koobi Fora by looking out the window of another single-engine airplane, when he was flying back from Ethiopia on a detour to avoid a storm. Once he returned to explore Koobi Fora on the ground, he and the museum's team of fossil hunters, known as the Hominid Gang, would soon uncover a stunning assortment of skulls and bones of early humans who had lived on the eastern shore of the lake 1 million to 2 million years ago. Those fossils would bring him so much publicity that by the age of thirty-four, in 1977, Richard had already appeared in *Life,* in *National Geographic,* and the front pages of all the major newspapers. He was the only paleoanthropologist of his generation to appear on the cover of *Time,* where an artist drew a hairy ape-man squatting beside Richard, as if the two were hanging out together in the Turkana desert. The headline was HOW MAN BECAME MAN.

But even as Richard and Meave worked on the eastern shore, they often gazed across the lake and wondered what lay in the fossil beds on the distant shore. Some of the ancient people they had found already had big brains and were using stone tools by 2 million years ago, and they were clearly well on their way toward becoming human. They were about the same age as Raymond Dart's Taung baby and other australopithecines discovered in South Africa, making the southern apes their contemporaries rather than their ancestors. Where were their forebears?

Richard and Meave knew about Kanapoi because Bryan Patterson, the Harvard paleontologist, had found the elbow of an early human there, almost twenty years earlier, in 1965. The discovery was tantalizing, since they knew by the 1980s that the elbow was at least 3 million years old and one of only a few scraps of bone from early humans of that era. Richard had visited the camp soon after the discovery. He saw the fossil as one poorly dated bit of bone, and lamented that Patterson's team had driven its vehicles over fossil sites. No one else had explored Kanapoi since that early expedition. But Richard always meant to return to redate the site and see if there were more fossils.

By the time Richard got back to Kanapoi, on the scouting trip with

Walker in 1981, there were still plenty of fossils at Kanapoi—but of animals, not human ancestors. The fossils the Hominid Gang found were broken, probably chewed by carnivores and trampled by the camels and goats that the local Turkana children herded along the deep gullies. The reconnaissance trip was cut short when Walker got ill and had to be flown back to Nairobi. Richard left Kenyan fossil hunter Kamoya Kimeu in charge of the camp. But soon after Richard's plane had taken off, bandits carrying rifles descended on their camp, demanding blankets and other camp supplies. Kamoya convinced the young men to leave and return the next day, when he would give them more supplies. The moment they left, Kamoya's crew hastily packed up the camp in the dark, stealthily loading up the Land Rovers and pulling up the guide ropes for the tents all at once, on a signal, before racing off into the desert without headlights.

"It was not promising," Leakey recalled, as he flew over Kanapoi with Meave a few years later. Although they were accustomed to working in hostile places, Kanapoi was not particularly compelling. Only traces remained of a vast lake that had once drowned the valley, now buried in sand and glazed with cobblestones and pebbles that made it difficult to spot small bones and teeth on the ground. A few lava-capped volcanoes rose out of the desert like black islands in a sea of brown sandstone. The only shade in the bleached lakebed came from an occasional spindly acacia that seemed to reach to the sky for water or a thatched *boma* erected by local tribesmen. Even the zebras and antelopes that were so abundant on the eastern shore of Turkana had abandoned this place. Clouds of dust whirled across the valley floor.

Richard and Meave didn't stop there. As director of the National Museums of Kenya, Richard had his pick of fossil sites to work on in Kenya. On that day the Leakeys were on their way back from seemingly more productive fossil beds farther north at Nariokotome, where they had painstakingly been working with Walker and the Hominid Gang to uncover the remarkably complete skeleton of a boy who had lived 1.6 million years ago. This ancient Turkana boy was the most complete skeleton of an ancient hominid that had been found—about 80 percent of the bones of this *Homo erectus* youth would be unearthed. It would keep them busy for some time. For now, they passed over Kanapoi.

⤬

Meave would remember Richard's prophecy about Kanapoi a decade later. It was the summer of 1994 when she got a call in Nairobi where she was working as head of paleontology at the National Museums of Kenya. It was Hominid Gang fossil hunter Kamoya Kimeu on the radiotelephone, calling from Kanapoi. "We have something for you," he told Meave.

Meave made the twelve-hour drive to Kanapoi as quickly as possible. She bounced along rutted roads that took her north along the edge of the Great Rift Valley, climbing gradually through lush farmlands where cattle grazed, sometimes alongside zebra, to the unexpectedly cool pine forests at the equator at Timboroa, where the altitude is sixty-eight hundred feet. The two-lane A-1 road then descends three thousand feet, twisting through the terraced farmlands of the green Cherangani Hills where the Pokot people sell baskets of tomatoes by the side of the road and bougainvillea climbs over their mudstone houses and thatched huts. The green farms eventually give way to the scorching Turkana plains, and the A-1 passes by scenery that is sometimes surreal: sisal branches rise in curlicues that look as though they could have been drawn by Dr. Seuss. Tiny dik-dik antelopes dart among the scrub brush and olive baboons scavenge for food in the trash thrown from cars. An overturned car or truck beside the pockmarked tarmac and bored-looking soldiers at checkpoints remind travelers of the dangers along the route, with its washed-out bridge, unmarked potholes the size of jeeps, and bandits armed with AK-47s.

By midafternoon, Meave reached the dusty market town of Lokichar, where barefoot little boys often surrounded her truck to beg for food. Here, Meave turned off the road to follow evanescent tracks in the desert, fording sand rivers in her four-wheel-drive Land Rover and shimmying up slippery banks. Sometimes when she drove this route she had to stop to put leaves from doum palms under her tires for traction, as local Turkana swathed only in sarongs of plaid cloth watched. Finally, Meave arrived at camp late in the day.

The Hominid Gang had set up tents under acacias along a dry streambed that cut right beneath spectacular cliffs with chalk-white layers of ash.

The next morning, as the soft dawn light tinged the rock-strewn hills a rosy hue, Meave drove out to the desolate gully where the Hominid Gang had been searching for fossils among the rubble of lava pebbles that glazed a gentle slope. Tall and fit, with long legs and a fast stride, Meave is known for setting the pace as younger team members scramble behind her over the hills. Her stamina is well known to the Hominid Gang—she eats little, braces herself with a cup of black tea in the morning, and often heads out alone for long walks over the arid terrain. She is a grandmother now, yet she retains a girlish enthusiasm for the fossils, exclaiming over an extinct Nile crocodile's "pretty face" as she extracts it from stone with a dental pick or calling out excitedly for other team members to come see a new hominid tooth. To observe her in the field is to see a woman who seems to draw strength from the desert around her—and who visibly relaxes, as if she is shedding the burdens that sometimes plague her in Nairobi, where being scrutinized as a member of the famous Leakey clan cannot be easy, especially for someone who prefers privacy. Where others wilt and fade, Meave is clearly in her element at Turkana.

On this summer day in 1994, Kamoya and the Hominid Gang took her to the spot where they had made a discovery that told Meave the team was finding a new type of hominid, never seen before. Hominid Gang associate Peter Nzube Mutiwa showed her three teeth camouflaged amid a carpet of lava pebbles. Meave carefully extracted them, and a year later in *National Geographic,* she described the moment when Kamoya cradled the teeth in his hands and almost whispered, "Surely this is where we came from." She knew what he meant: the teeth looked apelike but had a hint of something human about them. Like Kamoya, she felt awe—if the teeth really were 3.9 million to 4.1 million years old, as new argon-argon dates on the site indicated, they were significantly older than any other evidence of members of the human family then known. Could the teeth have belonged to a new species of hominid—even humanity's earliest known ancestor?

She congratulated Nzube. Then she and Kamoya marked out a large area on the hillside where the teeth were found, removed the bigger stones, and began using small brooms to gently sweep the dirt into dustpans for sifting through wire mesh in large trays. Sifting is backbreaking and tedious, but it is the only way to catch every last fragment of fossilized bone. The members of the Hominid Gang kept filling their trays and sorting through the gray dirt and pebbles from dawn to dusk, leaving behind conical piles of sifted sand that looked like giant anthills.

Over the next few weeks, the hard work paid off: the team members collected an almost complete set of the mysterious ape's lower teeth. They also found tooth fragments from another individual. Bit by bit they assembled critical pieces of evidence that a different type of early human had lived there as early as 4.1 million years ago—a new species that had never been seen before. "My hunch that Kanapoi would produce some remarkably early hominids seemed to be right," Meave would recall. For Meave, the discoveries at Kanapoi marked a turning point. At the age of fifty-two, she would finally get recognition as a scientist on her own merits, rather than as a member of Richard's team.

A zoologist by training in her native Wales, Meave Epps had wanted to do research on ships in the ocean, studying marine organisms. But in the early 1960s, every time she applied for a job on a research ship she was told, "Sorry, we're looking for a man." By the time she saw an advertisement in 1965 in the *Times* of London for a job to study monkeys at the Tigoni Primate Research Centre in Nairobi, she was ready to switch to animals that live on the land instead of in the sea. The man who interviewed her for the job in London was Louis Leakey. At the time, he was well known for launching Jane Goodall's study of chimpanzees. Less well known is that he also hired his future daughter-in-law Meave. She apparently impressed him with her ability to repair cars and care for animals; even though she was a surgeon's daughter who'd had a comfortable upbringing, attending convent and boarding schools, she had also cared for farm animals to earn extra money. Her job at the Tigoni center was to care for the monkeys and study monkey skeletons in the collection, which became the focus of her Ph.D. thesis in zoology.

She worked at the primate center for several years and after finishing her thesis became acting director at about the time she met Richard. At first, he asked her to teach him to dissect monkeys; eventually, he invited her to join his young team on his second field season in east Turkana, in 1969, to study the fossil monkeys. She describes that first season at Turkana and her impression of the lake as magical. "It's the lake. People have been there for millions of years—you feel this sense of continuity," says Meave.

The lake would also be the setting of many key events in Meave's life—she would return to Turkana year in and year out for three decades. She marked different stages of her life there every year, as a scientist, wife, and mother. The first photos show her examining fossils with Richard. Then she is there with her eight-week-old infant daughter, Louise, cooling her in a bin of water as she works on fossils at a table nearby. A few years later, Louise and her younger sister, Samira, work alongside the excavations of the Turkana boy, fetching water and delivering messages. A decade later, Meave would be photographed lying on the ground excavating a fossil with Louise, now a paleontologist with a Ph.D. herself, in charge of Koobi Fora.

The way that Meave and Richard worked at Turkana would also change over the years. In the early days at Koobi Fora and at other researchers' camps, the search for hominids was far from systematic. Researchers seldom picked a place to work because it was the right age. Instead, they went where the exceedingly rare fossils of extinct apes led them—working where geologists, local people, or scouting parties found fossils or stone tools that had eroded out of the ground, perhaps washed out by rains or disturbed by humans or animals. Once researchers found a hominid fossil or other significant remains of extinct animals, paleoanthropologists would then be able to get funding to bring in experts to use radiometric dating to tell them the age of the sediments and, by association, the fossils, and to explain the geology of how the fossils had been deposited in the fossil beds.

By 1989, when Meave began running the expeditions for the National Museums of Kenya, she and her colleagues had worked all over the vast Turkana basin for twenty years. Geologists had trekked over much of the terrain, mapping distinct layers of sediment, rocks, and ancient shorelines to figure out the geological history of the basin—how it was formed and reshaped over millions of years by water, volcanic eruptions, and the tectonic movements of the earth. When Meave was looking for a new site to work, she could be strategic and pick one of the spots on the western shore where geologists knew that the sediments—and fossils—were about 4 million years old and, therefore, older than Lucy's species. One of those places was Kanapoi. But when Meave set out on her own expedition in 1989, she opted to explore another place, one where Patterson had worked first—Lothagam Hill, a slab of rock that had been thrust upward, exposing sediments that were about 4 million to 8 million years old, the perfect age. Meave described it as an island of red rock in the sandy, windswept desert. Meave had flown over it many times, and she knew that Patterson had found a lower jaw from a hominid there in 1967.

From the beginning, the museum team found plenty of fossils of every sort of animal imaginable—extinct rhinoceroses, horses, hippopotamuses, giraffes, and saber-toothed cats, as well as fossil fish, turtles, crabs, and crocodiles. The team of specialists later filled a 678-page scholarly volume with detailed descriptions of these animals and set a high standard for describing the ancient environment where these animals lived. Those fossils would be used as index animals to help date other fossil sites in Africa, and to help reconstruct the migrations of mammals in and out of Africa more than 4 million years ago. But though the volume was entitled *Lothagam: The Dawn of Humanity in Eastern Africa*, one animal was virtually absent: the human kind. After five years of collecting, the team had only found six hominid teeth. The fossils seemed much rarer at sites earlier than 4 million years, perhaps because the populations of hominids were smaller or paleoanthropologists were looking in the wrong places—or both.

Meave was intrigued by the fossils of other animals, but she knew that without hominids, she would be unable to keep funding field expeditions at the site. She also wanted to know where the hominids actually were 4 million to 8 million years ago. She finally concluded that human ancestors living that long ago were probably not spending much time in the open swamps at Lothagam.

But before she could scout out a new site, her last season at Lothagam was interrupted by a plane crash. On June 2, 1993, just a week after camp was established at Lothagam, a plane flew in to bring Meave the news that Richard had been in a crash and was seriously injured. The single-engine plane he was flying had lost power and crashed. (Although some later suspected sabotage, no evidence was found of foul play.) Meave flew back to Nairobi immediately, and her daughter Louise, then a graduate student in paleontology, volunteered to run the camp so Meave could stay with Richard. Doctors were trying to save his injured legs, first in Nairobi, then in England. A renowned surgeon finally had to amputate both legs at the knee. Fitted with prostheses, Richard had to learn to walk again, and Meave was unable to return to Lothagam until late in the summer, when she helped wrap up the fieldwork there.

She thought again of Kanapoi and of making a fresh start. Meave and a small scouting party headed south late in the summer to check out Kanapoi, which had been virtually unexplored since Patterson's expedition. They found bone fragments of animals that were broken and chewed by ancient carnivores, but Meave felt optimistic enough to schedule a field mission to Kanapoi the following year. Later, geologist Craig Feibel collected samples of pumice that he and colleagues dated to 4.1 million years ago. This might be a better window of time to explore than the older layers at Lothagam, Meave thought. By 4.1 million years ago, hominids might have ventured out of the woods more frequently and into more open terrain, where their remains stood a better chance of being preserved.

Meave found that the deep gullies at Kanapoi included some of the most promising sites on Lake Turkana. Just two weeks after the museum's team had set up camp, Wambua Mangao called out excitedly. A veteran member of the Hominid Gang known for his huge hands and his talent for telling stories, Wambua had not picked up the fossils. Instead, he showed Meave five small areas of bluish tooth enamel that were embedded in rock. It was the upper jaw of an extinct ape the size of a chimpanzee. That night, Meave bought a goat from the local Turkana to roast for dinner, and the crew members drank beer to celebrate the discovery. A few days later Kamoya discovered a piece of shinbone. Then came the lower end of the same shinbone, but unfortunately not the middle piece. Still, the team was on a roll—and Meave knew, with a growing sense of excitement, that these were major discoveries, probably the remains of a new kind of early hominid. She reluctantly returned to Nairobi to attend to business.

Kamoya called her on the radiotelephone while she was working in Nairobi. The teeth that Nzube had found were the clincher. The canine was large, which suggested it was ancient. But other features, such as the shape of the root of the canines, looked more like those of members of the human family. She knew she had a hominid that was a lot like Lucy's species, but even more primitive. Could this be Lucy's ancestor?

Richard flew up to join the team during the last week of this remarkable field season at Kanapoi. He walked with artificial legs, but he was pleased to be in the field again, looking for fossils. While he and Meave were applying plaster to the skull of a large elephant, Nzube made yet another discovery. This time, he had found a complete lower jaw and next to it a piece of the ear region of a skull. As Richard excavated it, they watched. When it was finally out of the ground, they could see that these fossils showed the same mixture of ancestral chimplike features and more derived hominid traits found in Lucy's species. It wasn't clear if they were all members of the same species, but they were clearly significant discoveries.

Meave closed the camp and returned to Nairobi with these fossils, feeling thrilled with the summer's yield. She was transporting to the National Museums the most complete specimens of a hominid older than Lucy's species. She also knew that they probably belonged to a new species that would bump Lucy from her status as the earliest human forebear. Then she got another phone call. It was an American science writer. "Tim White has seventeen skeletons," she told Meave.

CHAPTER NINE

A VIEW FROM AFAR

While I was reading about the Afar, it occurred to me that if a rift valley is an ideal setting for preserving fossils and artifacts, then what about *three* rift valleys intersecting in one place? What about a depressed lowland that comprises one of the largest landlocked, multiple-rift-valley structures in the world?

JON KALB, geologist

When Tim White gives a tour of the Afar rift of Ethiopia, he often starts from space. He is selective about who he takes with him to the moonscape of the Afar, so the only way that most people will get to see it is to watch White's slide show. In lectures to the public, White starts his virtual tour by proposing that he take the audience up on the space shuttle to show them the view of the Horn of Africa. White shows satellite images of the Afar depression, a giant triangle of land on the northeast border of Ethiopia where it abuts the Red Sea and the Gulf of Aden. It is a view White knows well—for many years during the eight-year ban on fossil hunting in Ethiopia, these space-based images were just about the only glimpses he got of the Afar. White and his coworkers studied those images during the moratorium, even working with NASA scientists to learn how to translate different colors of land in the photos into various types of ancient sediments, such as basalt and exposed sandstone. The images would later be used as navigational tools to help them target promising areas of exposed rock.

Even to the untrained eye, it is clear from the space-based images that the Afar depression is different from the land on either side. The Afar is a low-lying wedge that has sunk below high ridges of darker land on either side, like a soft spot on an infant's head. But unlike an infant's skull, where the plates of bone will eventually move together and fuse, this depression is a continental crack where giant plates of the earth's crust are still spreading apart. It is one of the rare places in the world where three plates meet in a triple rift junction—and the center does not hold, exposing a giant rift zone in between as the plates pull apart in different directions. "What you're seeing here is the Arabian plate pulling away," White tells audiences, pointing to the northwest side of the Afar depression where it abuts the Red Sea.

Over 26 million years, the movement of the plates has pushed the Afar triple junction about 160 kilometers in a northeast direction. As the Arabian plate creeps off to the north in a counterclockwise direction at a rate of about 1.6 centimeters a year, it bumps up against the Eurasian plate and scrunches the terrain at that boundary into the Zagros Mountains. The Somali plate is pulling toward the south in a clockwise motion. And the giant African (or Nubian) plate is skidding to the northwest, colliding with the European plate, thrusting it up to create the Alps. If the plates continue on their trajectories, the Red Sea will become part of a new ocean that one day may engulf the Afar depression, though geologists debate whether the movement of the plates beneath the Afar has stalled.

The engine for all this tectonic activity is a hot spot twenty-nine hundred kilometers beneath the Afar triple junction, which is still burning a hole in the crust. A cauldron of extremely hot magma began pushing up from the earth's mantle to the surface more than 30 million years ago, forming a vast dome that stretched from Ethiopia to Yemen. At the peak of the Afro-Arabian dome was land that is now Ethiopia, which remains a high plateau and the most mountainous nation in Africa. One leading scenario for the formation of the rift zones is that molten magma pushed up under the dome, finally bursting through weak spots in the earth's crust. When the dome broke, it split along the boundaries between the plates, forming cracks that are the great rift zones, including the Afar depression and the Great Rift Valley.

Over millions of years, the energy at the hot spot has been released in a flurry of erupting volcanoes that spewed lava, ash, and rock in the rift zones. Over time, the valley floor sunk and the lava and volcanic material from the mantle oozed up onto the floor of the rift zone. As the two sides of the rift spread apart, the cooling lava spread out onto plateaus that look like ripples of sediment frozen in time. The hot spot is still fueling volcanoes that, along with tectonic movements, have inadvertently done the groundwork for White and other paleoanthropologists—they have exposed ancient sediments that would otherwise be buried, in the same way that a road crew uncovers walls of layered rock and dirt as they cut through a hillside.

The tectonic activity is not the only reason the Afar depression is the ideal place for finding fossils. "The cloud deck is important," says White, as he shows a slide of the Afar visible through banks of clouds. "You have to think about the water that comes when these clouds produce rain." The water falls down onto the giant dome and into the fissures of the rifts and the low-lying Afar depression.

The Afar, like other rift valleys, is a giant basin. Water flowing down rivers, tributaries, and drainage ditches in the rainy season is thick with dirt, rocks, and mud, which get dumped on the low-lying ground at the bottom of rift valley lakes and floodplains. The water and sediment are essential for turning bones into petrified rock, or fossils. If an extinct animal or human ancestor dies on dry land, its carcass is usually devoured by carnivores and its bones weather and decay into dust.

But if a hominid is buried quickly and gently by sediments at the shore of a lake or riverbank, for example, it might become the rare creature whose bones fossilize instead of decomposing. Once it's been buried by fine sediments, there is little oxygen, heat, or moisture to decompose the bones. In ideal circumstances, the bone doesn't disintegrate, but its organic material, such as DNA and proteins, disappears over time. The remaining spaces or pits in the bone and teeth absorb hard minerals, such as calcium and silica, which percolate slowly into the bone from water in the soil. Over time, the bone fills with hard minerals and becomes petrified, turning into stone.

While the entire eastern African rift has been called the cradle of mankind, a more apt analogy would be to call it the graveyard for humankind. Hominids ranged far beyond the Great Rift Valley, but the acidic soils in tropical forests probably disintegrated their bones before they could fossilize, and much of the rest of the continent's surface has eroded away. By contrast, the Afar valley was a trap for fossils.

White zooms in on one of the best dumping grounds for bones in the Afar—an area of exposed sediments 150 kilometers south of the triple junction on the western margin of the Middle Awash where he and his colleagues have focused their quest. "Let's go down on the ground and see what's happening there on our trip through time," says White, as he shifts to photos taken on the ground.

At Aramis, a drainage basin just west of the Awash River, a volcanic dome rises sixteen hundred feet above the valley floor. White shows how the slopes of the dome are cut in a series of stepwise ridges that climb up at a slant from the valley floor. Layers of rock and sediment are stacked up on the slopes, from the oldest, at 5.6 million years, at the bottom to the youngest at the top. But all the layers are not in one place—different ridges show different slices of time, with some overlap. By tracing these layers horizontally, sometimes for several kilometers, the team's coleader, geologist Giday WoldeGabriel, can link the layers of sediments to see where one period of time ends and the next begins. If all of these layers of exposed rock were stacked up from oldest to youngest, they would form a vertical layer cake one thousand feet high, almost the height of the Empire State Building.

Each site offers different windows back in time, with some overlap in the views into the past. At Aramis, White and WoldeGabriel were lucky enough to find a distinct white ribbon of ash that is the signature of a volcanic eruption 4.4 million years ago, which was laid down directly on top of a layer of salmon-colored sediment that was deposited by floodwaters that filled the basin for decades or even centuries immediately before that eruption—and that is filled with fragmentary fossils. Team members don't just trust their eyes to trace and map those layers—WoldeGabriel and others take samples of the ancient soil and rock back to their labs in the United States to identify the unique chemical signatures of different sediments of different ages. The glass shards and crystals were formed under intense heat and pressure in different volcanic eruptions, and each has unique chemical fingerprints that help geologists on the team identify and date the different layers of sediment. Using an electron microprobe in their laboratories, they can tease out the ratios of silica, titanium, aluminum, iron, magnesium, and other elements that are the distinct fingerprints of different volcanic eruptions. This identification and naming of different layers of sediment is known as tephrastratigraphy, and is critical for tying fossils to distinct strata of dirt so they can be dated.

The volcanic crystals in these volcanic horizons, called feldspars, also are packets of data that can be heated with lasers in the lab to release the gases trapped inside. The ratio of two gases in particular—two isotopes of argon—in a single crystal can tell how long ago the feldspar was formed, and so date the volcanic eruption.

As White flashes an image of Aramis, the bleached landscape is so blindingly bright that even a slide of it makes members of the audience blink. White jokes that to re-create the real conditions of the Afar, he would have to close the doors of the auditorium where he is speaking, heat it to one hundred degrees, blow in dust and sand, and bring in two dump trucks filled with scorpions, snakes, and malarial mosquitoes. Then, all joking aside, White notes that the working conditions are incredibly harsh, and that it takes a unique team of intrepid and dedicated researchers to return to the Middle Awash year after year. But he has never had any trouble recruiting students or colleagues. He shifts the focus of the talk from the geology of the Afar to his colleagues, showing members of the team working in front of a nearby volcano. He zooms in on several Ethiopians, including the first Ethiopian scientists to join the Middle Awash group, biological anthropologist Berhane Asfaw and geologist WoldeGabriel.

⌇⌇⌇⌇

In November 1989, Berhane Asfaw invited Tim White, Giday Wolde-Gabriel, and several friends to accompany him on an excursion to survey the Ethiopian rift for promising fossil sites. With a grant of $29,000 from the National Science Foundation and a permit from the Ethiopian government, they drove two Toyota four-wheelers down steep drainage canyons feeding the Great Rift Valley, in search of fossil sites. They also invited a half dozen young Ethiopians, notably Yohannes Haile-Selassie, who would become a formidable fossil hunter, and Berkeley graduate student Gen Suwa. Using NASA satellite photographs, topographic maps, and aerial photos, they honed in on promising areas of eroded sediments, driving along dry riverbeds and stopping to talk with the local people, some of whom had never seen Westerners before. Over the next few years, they found many promising areas with sediments of different ages, in addition to mapping many fossil sites they discovered. They were not allowed to collect fossils, but by the time the moratorium was lifted in August 1990, they were well positioned to pick up the thread of fieldwork in the Middle Awash that had been lost in 1982.

They also were a well-trained team that had slowly come together over the course of a decade. Starting in 1979, Berkeley archaeologist Desmond Clark had recruited Berhane Asfaw, who was a student majoring in geology at Addis Ababa University. Clark had read a paper Asfaw had written summarizing Ethiopian prehistory for the Ethiopian Ministry of Culture, and he had asked Asfaw's professors to introduce him to Asfaw. Then twenty-eight, Asfaw had a summer job at the Ministry of Culture and was fascinated by his nation's rich cultural and anthropological heritage. By the time he met Clark, he was eager to learn more.

Clark mailed Asfaw an application to Berkeley's graduate school. Asfaw applied, was accepted, and enrolled in African prehistory. He was at home in Clark's lab. It was a mini Pan-African congress of sorts—there was one student from Zambia, one from Nigeria, two from Malawi. One student from

Kenya had just left. "Basically, it was Desmond's dream to have lots of Africans involved in African prehistory," says Asfaw. "He realized the value of training local people." Asfaw joined Clark and White's field expedition to the Middle Awash in 1981, and soon became infected with fossil fever.

Asfaw also introduced Clark and White to WoldeGabriel, who was a lecturer at Addis Ababa University. By the time he was invited to go to the Middle Awash, he was thirty-six years old and had already overcome many obstacles to earn his master's degree in geology. He was born in northern Ethiopia, in Tigray, where his father, who had been a subsistence farmer, died when he was young, leaving his mother to rear five children by herself. Wolde-Gabriel was one of the poorest children at school, but he excelled in his studies and won a scholarship to boarding school. He entered Haile-Selassie University (now Addis Ababa University), and met Asfaw in 1974 when both joined the geology department. Their friendship persisted even when the university was shut down in 1975 after the overthrow of Emperor Haile-Selassie and students were dispersed to the countryside to work in mandatory civil service jobs. WoldeGabriel spent two years in southern Ethiopia building clinics and schools. After he completed his service, he returned to Addis Ababa, where he was one of only three geology students to return out of a group of eighteen who had been in the department before the university shut down.

WoldeGabriel's connections to Asfaw and the Berkeley group led to an offer for him to move to Cleveland to earn a Ph.D. in geology with Jim Aronson, a well-regarded geologist then at Case Western Reserve University. Aronson, now at Dartmouth University, was dating the site where Lucy was found at Hadar. WoldeGabriel earned a Ph.D. in 1987, and landed a job at the Los Alamos National Laboratory in New Mexico, where his specialty is studying the geochemistry of volcanic rocks and the formation of rift systems.

Meanwhile, Asfaw had also completed his Ph.D. in biological anthropology in 1988 at Berkeley, and his wife had earned an M.B.A. at a college in San Francisco. They returned to Ethiopia with their new degrees, fired up to do work in their own country. But there were few jobs, no housing, and the eight-year ban on fossil hunting was still in effect. Asfaw built his family a small house. He also approached the Ethiopian minister of culture and suggested that if the government was rewriting regulations for antiquities research, it would be useful to know the location of promising fossil sites so they could be regulated properly. Asfaw got permission to survey the Ethiopian rift, and he called the project the Paleoanthropological Inventory of Ethiopia.

Soon after Asfaw and his colleagues completed the survey, the director of the National Museum in Addis Ababa retired. In 1990, Asfaw became interim director of the museum and the NSF lab that had been built in 1981—a position he would hold until 1992. Also in 1990, the Berkeley-Ethiopian team and Donald Johanson's Hadar team were among the first groups of foreigners in eight years to get permits to return to Ethiopia to search for fossils (not just to survey sites). It was an optimistic time for paleoanthropology in Ethiopia. Several teams, including Johanson's group from the newly formed nonprofit research center that he had opened in Berkeley—called the Institute of Human Origins—returned to Ethiopia in 1990. The researchers from different teams were still on friendly terms, still cooperating with one another and showing one another the fossils they were finding. They had gotten a fresh start in Ethiopia, and John Yellen, program director for archaeology at the National Science Foundation, congratulated the teams for improving relations with the Ethiopian government. He singled out Clark and White, in particular, saying that they deserved a lot of credit for training Ethiopians who were returning to Ethiopia to assume important positions there.

Everything came together for the nascent Middle Awash group in that first season. As they prepared to return to the Middle Awash, they reactivated Clark's grant and used the remaining funds to revisit the areas where they had begun exploration a decade earlier.

CHAPTER TEN

THE ROOT APE

The creature widely named as the Missing Link was seen within the
community more as cause for quiet satisfaction than wild consternation.
There was, indeed, a morphological gap between apes and known
hominids that was waiting to be filled, and the hominid from Aramis in
Ethiopia fitted the bill nicely. The easy placement of *A. ramidus* in the
human family tree was a sign of maturity in a field usually distinguished
by discord. Famous last words?

HENRY GEE, associate editor, *Nature*, 1995

On June 10, 1994, an editor at the prestigious British scientific journal *Nature* returned from lunch to find a Post-it message stuck to his desk that
simply said, "Mr. Asfaw called." The editor, Henry Gee, quickly dialed the
number, hoping it was Berhane Asfaw with news of some mysterious new
fossils from Ethiopia. It was indeed Asfaw, who was traveling from Berkeley
to Ethiopia and had stopped over in London for the night. Would Gee meet
him at his hotel in South Kensington right away?

Half an hour later, Gee was waiting in the lobby of a modest hotel near
the Natural History Museum. The elevator door opened and Asfaw emerged,
carrying a large brown envelope and with a mischievous smile on his face.

Gee took Asfaw around the corner to the basement bar of the Norfolk Hotel, which was a favorite hangout for researchers at the museum. Once there, Asfaw quietly handed the envelope to Gee. It contained documents and photographs describing seventeen specimens in the safes at the National Museum in Addis Ababa. Gee, an associate editor at the journal *Nature,* would joke later that he felt like a character out of a John le Carré novel, receiving state secrets. (In truth, Asfaw, passing through London on his way from Berkeley to Addis Ababa, thought it would be faster to deliver them by hand than by mail.) In fact, the photos that Asfaw passed to Gee were revolutionary—they revealed the identity of a long lost ancestor for humankind. The photos showed the fossils of a new kind of hominid never seen before—one that was alive 4.4 million years ago, which was almost a million years earlier than the oldest members of Lucy's species.

The new fossils came from Aramis, just seventy-five kilometers south of Hadar, where Lucy's skeleton had been found twenty years earlier. They had been discovered by members of the Middle Awash Research Project. What Gee saw in the photos was not seventeen skeletons, as Meave Leakey soon would hear. Instead, Gee saw seventeen fossils. Although they were mostly of teeth and jaw fragments from seventeen different individuals, he was not disappointed. He and Asfaw toasted the discovery with bottles of Grolsch beer. Gee left delighted that he had gotten the manuscripts announcing the first new species of hominid since Lucy's baptism as *A. afarensis* in 1978.

He had been waiting to get manuscripts from this team for more than a year, ever since he had met with Tim White in Berkeley the year before, on February 14, 1993. Gee had gone to see White in his lab at Berkeley and had gotten an unexpected Valentine's Day gift. They talked mostly about White's research in detecting signs of cannibalism on fossil humans in Europe and Africa, and on the Anasazi Indians of the southwestern United States. Only at the end of the discussion did White mention casually that he had new photos of fossils discovered during the last field season in Ethiopia. He tantalized Gee with a set of monochrome contact prints, which included bone fragments that looked more like cornflakes to Gee than impressive fossils of jawbones and teeth. As Gee was trying to make sense of the photos, White said, "This is the earliest known hominid." He said it with self-deprecating humor, in a manner that Gee thought was proud but that also acknowledged how unimpressive was the fragmentary evidence for a hominid of such great antiquity and importance. White also told Gee he would give him a manuscript in a year and a half.

Now, sixteen months later, Gee had two manuscripts—one by White, Asfaw, and former graduate student Gen Suwa that described the fossils, and a second one by a larger team led by WoldeGabriel that described the fossil site and the ancient environment where the hominid had once lived. The photos that had been difficult to parse earlier now clearly showed a jaw fragment, the base of a skull, three arm bones, and some teeth. The authors were proposing that those fossils were the remains of the earliest known hominid, which was alive 4.4 million years ago. It was the first time that a new species had been proposed as the earliest member of the human family since 1974, when Johanson and his colleagues unearthed Lucy.

The new hominid was even more apelike than Lucy, showing a mix of primitive and more evolved features, as was expected for earliest members of the human family. Its skull bone and one milk molar, in particular, resembled those of a chimpanzee more than the teeth of Lucy and other hominids. But it did not appear to be an ancestor of chimpanzees, because its evolved features tied it to later hominids rather than to chimpanzees, suggesting that it was a distant ancestor of hominids like Lucy. It made sense that it looked more primitive than Lucy, since it was alive 1.2 million years—or eighty thousand generations—before Lucy and 800,000 years before the earliest members of her species, *A. afarensis*. The seventeen fossils were also 300,000 years older than the new hominid Meave Leakey and the Hominid Gang were unearthing at Kanapoi that same summer, even as they were unaware that White and his colleagues had found an even older candidate for earliest human ancestor.

Meave Leakey was not the only one to be surprised by the news: almost no one knew that the Middle Awash team had discovered this new type of human ancestor in Ethiopia two years earlier, in the fall of 1992. Rumors would begin to surface only in the summer of 1994, as a few scientists were asked by Gee to review the manuscripts for *Nature* and word inevitably slipped out that fossils of a new type of human ancestor had been discovered. In fact, with the discipline that would become characteristic of this team, White and Asfaw and their colleagues had kept the discovery under wraps for eighteen months, ever since Suwa had discovered the first of the seventeen fossils at a new site they were exploring near Aramis.

The members of the Middle Awash research project had spent weeks that year exploring another area, at first concentrating on barren outcrops where they had found fossils of animals that had lived more than 5 million years ago. But they had found no hominids. The season was ending soon, and the absence of mammals of the human kind was beginning to bother them. Working in the Middle Awash had not gotten off to a smooth start—in their first field season after the ban had been lifted one year earlier, in 1991, they had been working on the eastern side of the Awash River when they got caught in a conflict between the Afar and the bordering Issa tribe and gunshots hit their fuel barrels. They aborted their research.

When they returned the next year, in 1992, they pushed west of the Awash River and began to work on sites of many ages. They returned to a place they had visited in 1981, upstream of the small village of Aramis. It was inauspiciously named after a dry streambed that was part of a latticework of sand rivers that drained the muddy runoff from the fossil-bearing badlands into the Awash River during the rainy seasons.

There, on the morning of December 17, White, Asfaw, Suwa, and several other team members spread out on foot, searching for fossils among the multitude of pebbles paving the desert floor. They found lots of fossil wood, and some fragmentary fossils of animals that were not particularly significant but that included many monkeys. As the midday sun bore down on them, some of the team members were beginning to tire and think about stopping for lunch. Then Suwa saw the glint of something shiny among the pebbles of the desert pavement. It was the polished surface of a molar. "I knew immediately that it was a hominid," Suwa would say later, explaining that the shape of the tooth looked more like that of a hominid than an ape. At the time, he merely announced to other members of the team: "I've got one."

Suwa, now a paleoanthropologist at the University Museum, the University of Tokyo, in Japan, knew the molar was older than Lucy's species because the team had found other animal fossils in the same layer of sediment earlier that day that were more than 4 million years old. "I knew it was one of the oldest hominid teeth ever found," he says. But there was no whooping or shouting or jubilant celebrating in camp that night. Instead, the team reacted with a sense of relief and pride that their hard work had resulted in someone finding a hominid. They resolved to renew their efforts—White had made it clear to all of them that time was precious in the field. The excitement would come later.

The next morning, the team got back to work. It was the end of the field season in the Middle Awash, and they would have to use their time wisely to recover every bit of hominid bone they could find in the few days they had left. They scoured the area on foot, sometimes crawling on hands and knees, working from dawn to dusk, with only an hour off for lunch. Suwa found more teeth; paleoanthropologist Scott Simpson of Case Western Reserve University found an arm bone and a bit of skull bone. They gently used brooms to sweep up buckets of dirt, then poured the dirt into giant screens so they could sift it for every bit of tooth and bone. As they assembled bits and pieces of fossilized bone, they were working so intensely that they barely had time to process what they were finding—until a former Ethiopian antiquities representative named Alemayehu Asfaw found a jaw. Alemayehu Asfaw clearly had a knack for finding jaws—he had spied the first jawbones of Lucy's species at Hadar in 1974, the same remarkable year that Lucy was found. This time, he found the lower right jaw of a child, with the milk molar still attached. When Suwa saw the jaw, he said with classic understatement, "I think this is pretty important." The tiny blue-gray milk molar attached to the jaw fragment resembled a chimpanzee tooth. That's when the individual team members felt the thrill of discovery; they recognized that the milk molar was so primitive that it belonged to a species that was older and more primitive than Lucy's species, *A. afarensis*. White would say later that the deciduous (or baby) molar from the child's jaw was a complete surprise. Its shape was so different from the same milk molar in later human ancestors that it alone gave them enough evidence to know they were looking at the partial jaw of a new species. But they didn't think they had found an ancestor of chimpanzees, because the jaw also held a lower incisor that had not erupted that was decidedly similar in size and shape to the same tooth in hominids. Other fossils from Aramis also bore evolved traits shared only with later hominids.

The team returned the following year, at the end of 1993. On December 29, 1993, a local Afar man from Aramis, Gada Hamed, found a set of ten teeth from one individual. Hamed would die tragically a few years later, shot by a member of the Issa tribe in a battle on the other side of the Awash River. But he would be honored as the discoverer of the tooth that would become the type specimen for the new, as yet unnamed hominid.

It would be a remarkable field season, and the team's success in finding fossils would also prove White's and his Ethiopian coleaders' skill in pulling together, funding, and managing a first-rate coalition of students and researchers. They came from all over the world, from disparate backgrounds— Japanese and Ethiopian students, some from humble origins, working on their hands and knees alongside White and even Ann Getty, the wife of San Francisco billionaire Gordon Getty. At the time, she was a student studying anthropology at Berkeley, where she had taken White's courses. She and her husband had long been leading donors to the nonprofit organizations that fund anthropological research, but now she could take an active part in the research herself. She and Gordon Getty flew the team from San Francisco to Addis Ababa in their private jet. Ann Getty would contribute more than logistical and financial help to the team and the National Museum of Ethiopia: White is probably the only paleoanthropologist who has thanked the wife of a billionaire in the footnotes of a scientific manuscript for the fieldwork she actually did and for discovering fossils.

After the field season ended in 1993, White and other members of the team returned to Addis Ababa, where they cleaned the fossils and compared them directly with collections of Lucy's species, *A. afarensis*. They made casts of the teeth and bones, since the new Ethiopian laws required that the original fossils stay in the museum in Addis Ababa, as did similar laws in Kenya and most other nations. The days of signing out fossils, such as Lucy's skeleton, to take home to the United States for study had passed. White and Asfaw had been among the researchers who were alarmed when fossils taken out of Ethiopia in the 1970s were returned, some never examined and still in crates that had never been opened. Many fossils were broken and damaged during transport. The new regulations explicitly stated that no fossils were to be removed from Ethiopia. Team members would take detailed measurements, photographs, and X-rays instead. They would use those data and casts to compare the new fossils with the same teeth and bones of other hominids and extinct and living apes in museum and university collections in Kenya, South Africa, Europe, and the United States.

The point was to find shared derived traits, novel evolutionary characters that the new fossils and their apparent descendants, the australopithecines, had acquired after we split from the ancestor we shared with chimpanzees. The researchers were looking for novelties that had arisen as exclusive features in the hominid lineage after it had diverged from the chimpanzee lineage and that were not primitive traits passed on from ape ancestors.

They found what they were looking for: derived traits in the teeth and fragments in the base of the skull that were shared with australopithecines, but not chimpanzees or gorillas. Two traits, in particular, suggested that the specimen was a member of the human family, rather than a member of the chimpanzee family. It had an upper canine shaped like an incisor or a diamond from the side view, rather than the triangular, daggerlike point of a chimpanzee's canine. The new canines also did not rub, or hone, against the premolar below during chewing. The researchers did not have direct evidence that the new species walked upright, which was the traditional marker of being a hominid. But they did have indirect evidence from skull fragments that suggested that the creature had a shorter skull base, as seen in later hominids that walked upright.

They also found that the new fossils retained many primitive traits that were chimplike, which showed that it was not a member of Lucy's species, *A. afarensis*. In the case of the surprising deciduous molar, for example, they compared it with the same tooth in other species of hominids and apes, ranging from modern humans to australopithecines and two species of chimpanzee. They found that it fell well within the range for chimpanzees' baby molars in every dimension, illustrating clearly how primitive and apelike the original owner of that molar was in its development of milk teeth.

After doing similar comparative analysis for all the teeth and fossil bones, the researchers concluded that they had a new species that was neither chimpanzee nor *Australopithecus afarensis*. The next step was whether to classify the fossils as members of *Australopithecus* or to put them into a new genus of hominid. At the same time, White reported it was "the most apelike hominid ancestor known." The team members were initially conservative and put it in *Australopithecus*, but they added the caveat that they anticipated finding more fossils of lower limbs that might lead them to put it in a new genus of hominid later. This stunning fossil was from "a long-sought potential root species for the Hominidae," White and his colleagues wrote in *Nature*. They provisionally baptized the new species *Australopithecus ramidus*, drawing on the Afar word *ramid*, or "root," and submitted their papers to *Nature* in June 1994.

Three months later, on September 22, 1994, the cover of *Nature* featured the baby molar set in the jaw like a diamond mounted in a setting from which a few other gems had fallen out—and held up like a priceless ring between a man's thumb and two fingers. The red headline underneath said simply, EARLIEST HOMINIDS.

⸻

Lucy's status as earliest hominid had been usurped by *A. ramidus*. The reaction to the ascension of *A. ramidus* to Lucy's place of honor as matriarch of the human family was exuberant. An accompanying article that ran in *Nature* under the headline THE OLDEST HOMINID YET set the tone. Paleoanthropologist Bernard Wood, then at the University of Liverpool, wrote that the new fossils would push back humans' knowledge of their own lineage by more than half a million years. The root ape was recognized as a long lost member of the human family, slipping nicely into the slot between apes and known hominids on the family tree. Encouraged by the editors at *Nature*, Wood invoked a term that had long since fallen out of favor, concluding,

"The metaphor of a 'missing link' has often been misused, but it is a suitable epithet for the hominid from Aramis."

Nature showed little restraint in its promotion of the seventeen fossils from Aramis as "a missing link," and newspaper and magazine headline writers took the additional step of ignoring the qualifying article "a." The news made the front pages of newspapers in London, New York, San Francisco, and around the world. The headline in the *Times* of London was typical, calling it THE BONE THAT REWRITES THE HISTORY OF MAN. More than one article said that the new fossils supported the view that there was a single line of descent from chimpanzees to humans, with diagrams of family trees showing *A. ramidus* as the direct ancestor of Lucy's species, *A. afarensis,* and giving rise to all the different types of hominids that came later. It was pleasing in its clarity and logic. Lucy's discoverer, Donald Johanson, would be quoted in the *Los Angeles Times* as saying that beyond 4 million years ago there appears to be "a single line and a single lineage." Even White would invoke the image of a missing link, saying that this new species of hominid was the "oldest known link in the evolutionary chain that connects us to the common ancestor with the living African apes. The discovery takes us one major step closer to this common ancestor."

Those kinds of comments would prompt Henry Gee at *Nature* to eventually regret encouraging the use of the term "missing link." In his book *In Search of Deep Time,* he wrote that the term initially seemed more palatable than phrases such as "the hominid closest to the evolutionary split between our lineage and that of apes." He explained, "As editors of *Nature,* we were, on reflection, wrong to pander to the voodoo paleontology as portrayed by the media, because it presupposes a model of evolution that is linear, upwards, and progressive. We know that this model is mistaken, and yet it is deceptively easy to see evolution in this way, especially when we are discussing our own origins."

The paleoanthropological community nonetheless welcomed the new fossils in 1994 with what Gee described as "quiet satisfaction," because they fit nicely into a gap that was waiting to be filled. They met the expectations of many researchers of how that ancestor should look. Wood had welcomed the new fossils as an expected arrival, noting that "it is a sign of growing maturity of paleoanthropological research that, important as the discovery of *A. ramidus* is, the presence of a hominid much like it had been predicted." For several months, paleoanthropologists could revel in the certainty that the new fossils fit their notions of the earliest stages of human evolution.

In truth, the real importance of the new fossils was that they offered the first real look at the anatomy of an extinct ape or human ancestor of any kind from the shadowy period before 4 million years ago. These seventeen fossils represented almost the entire fossil record for apes and hominids from the crucial time 4 million to 7 million years ago. There were still no traces of the ancestors of chimpanzees or gorillas. The only other fossils of hominids older than 4 million years were the elbow and fragment of jaw that Bryan Patterson found in west Turkana and the partial jaw, arm bone, and molar that Andrew Hill and Martin Pickford found in the Tugen Hills. Meave Leakey and Alan Walker had yet to announce their slightly younger discoveries at Kanapoi and Allia Bay in Turkana.

The Aramis fossils also showed that early hominids are not chimpanzees. Researchers have often used chimpanzees and gorillas as surrogates for the earliest human ancestors over the years. Indeed, Pilbeam and Harvard primatologist Richard Wrangham, who studies chimpanzees in Uganda, have proposed that the earliest human was probably chimplike. They have noted that gorillas are so similar to the two living species of chimpanzees—the common chimpanzee and the bonobo—that they are often described as scaled-up chimpanzees (except for in the dentition). The simplest explanation for those similarities, such as knuckle-walking, where they put their forward weight on their knuckles, is that they inherited them from their common ancestor. But this common ancestor would also have to have been the ancestor of humans, because the ancestor of gorillas split off from the line leading to humans before the ancestor of chimpanzees did. Pilbeam and Wrangham have sketched a rough profile of the common ancestor of humans and chimpanzees (after the split with the ancestor of gorillas). The skull is the shape of a chimpanzee's, the teeth are covered in thin tooth enamel, and, in their opinion, its hands were used for knuckle-walking. If the last common ancestor of humans and the African apes looked a lot like a chimpanzee, then it is most likely that the earliest members of the human family also did. Indeed, the team found that the new root ape did share many traits with chimpanzees. White would stress, however, that *A. ramidus* was *not* a chimpanzee—and that many of the traits seen in modern chimpanzees and gorillas arose after they split from the line leading to humans. The new fossil hominid had a shorter face and its non-honing canine was less fanglike, suggesting that the males in this species may have already adopted a different social strategy for competing for mates—one that involved less fighting over fertile females, for example, than in modern chimpanzees.

It wasn't just the anatomy of *A. ramidus* that offered surprises. The second paper in *Nature*, written by geologist WoldeGabriel and his colleagues, offered a glimpse into the ancient world inhabited by *A. ramidus*—and it looked like a forest. The landscape at Aramis 4.4 million years ago was very different from the hardscrabble badlands there today. *A. ramidus* died in the woods of a tropical floodplain that was teeming with forest-dwelling kudu antelopes, colobine monkeys, birds, bats, rodents, and even otters. The dominance of colobine monkeys and kudu, in particular, in the same layer of sediment as the hominid fossils was a strong indication that the hominid died in a closed wooded environment, WoldeGabriel wrote in *Nature*. They also found abundant petrified wood, fossilized tree seeds, and even millipedes, which all pointed to the woods, not the open savanna. Indeed, seeds from a plant that bears fruit that chimpanzees eat not as a staple but as a fallback food today suggested that *A. ramidus* lived in the type of habitat where chimpanzees could also have survived, even though none were found. Chew marks on the bones indicated that carnivores had feasted on some of the remains—and, in fact, the remains of large cats, hyenas, and bears were among the fossils discovered.

Just as notable were the species of large mammals that were missing. Rhinoceroses, horses, and antelopes similar to gazelles and oryx, which roam open grasslands, were rare at Aramis. This indicated that *A. ramidus* probably still lurked under the cover of the trees, where fruit and other food were to be found. There were some species that complicated the picture—one species of burrowing rodent that is not usually found in forests was also detected at Aramis. Regardless, it was not the grassy savanna that most paleoanthropologists had envisioned when they thought about the time and place where the earliest hominid stood up and began walking upright to peer over the high grass, perhaps in search of game or to avoid carnivores.

The notion that upright walking—and therefore the transition from ape to human—had evolved in the open country had persisted ever since Raymond Dart realized that the Taung baby had lived in a nearly treeless veldt in South Africa. A generation of researchers came of age influenced by the savanna hypothesis, which assumed that all the key phases of human evolution took place in the open grasslands of Africa. Once human ancestors emerged from the cover of the trees, they adopted upright walking so they could scavenge or hunt game better in the open grassland than if they were moving on all fours. Walking upright meant they could see farther, carry tools and food more efficiently, hoist weapons, and intimidate predators with their large stance.

This view came entirely from fossils of hominids that were alive in the past 2.5 million to 3 million years. Lucy's species, *A. afarensis,* also came from open habitats, but new studies would soon find that the earliest australopithecines still favored well-wooded areas near the shores of lakes and rivers. Perhaps *A. afarensis* was one of the earliest hominids that could adapt to a wide range of habitats, perhaps retreating to trees to sleep at night or as refuge from harm. The notion that the cradle of humankind was a grassy savanna was clearly wrong if *A. ramidus* was indeed a hominid. WoldeGabriel's paper ended with an interesting observation: one explanation for the difficulty researchers had finding the fossils of the earliest hominids might be that they had been looking in the right time, but the wrong place.

There was still one stumbling block preventing many scientists from embracing this revisionist view that hominids arose under cover, among the shady boughs of trees. Many still wanted direct evidence that *A. ramidus* had walked upright or was at least evolving the ability to walk on two legs before they accepted the species unequivocally as a hominid. The identification of *A. ramidus* as a hominid rested primarily on its teeth, which brought back memories of the misidentification of *Ramapithecus* and *Kenyapithecus* as hominids. Indeed, anatomist Elwyn Simons, the longtime champion of *Rama-pithecus* as a hominid, recalled that when *Ramapithecus* was known only from jaw fragments, he had also drawn up a list of a dozen or more derived traits that it shared with *Australopithecus*. He had warned that any identification of a species' identity based on its teeth alone was unlikely to be as accurate as an identification based on parts of the skull and the skeleton below the neck as well. Unlike Rama's ape, the new root ape was not a "dental hominid," because its identification also relied on the base of the skull and the shape of the arm bones.

There was one obvious way to settle the question. Find hip or leg bones for *A. ramidus* to determine once and for all how it walked. Indeed, White would say that he would be delighted with a thigh—and while the team was at it, it would be nice if somebody would find an intact skull as well. Easily said, but not easily done. "We have only begun to understand this ancestor and its environment," he admitted. "I am sure there are more surprises in store for us."

No sooner had White wished out loud for a femur and a skull than the Middle Awash group delivered. It was uncanny. Less than two months after the article in *Nature* announced the discovery of *A. ramidus*, in November 1994, immediately after the rainy season had stopped, the team returned to Aramis. On the first day, November 5, they were crawling along on hands and knees less than two hundred feet from where Gada Hamed had found the type specimen the year before, when Ethiopian Yohannes Haile-Selassie spotted two fragments of a hand bone. Haile-Selassie, who was a graduate student at Berkeley at that time, called to Berhane Asfaw, who was nearby. Asfaw thought it was an animal bone, but Haile-Selassie knew it was a hominid hand bone. The two were worried because it was so light—it was in such poor, brittle condition that they wondered if it was a bit of modern bone that had not fossilized yet. They discussed it for a few minutes, and then sought White's advice. White took a look at the bone and said that Haile-Selassie was dead right. It was a bone from the palm of a hominid's hand and it was extremely old. But by now it was late in the afternoon and the light was fading, so they returned to camp with only the two fragments.

Over the course of the next few days, they swept up the dirt in the area and sifted bucket after bucket of rubble through a sieve. The tedious sifting and thorough scraping of the ground surface paid off—they found more hand and foot bones. White described the fossils they were finding to Henry Gee at *Nature,* who was writing a New Year's article summing up new additions to the human family tree. Gee was tempted to call it a partial skeleton. But White was reluctant. Partial skeletons are incredibly rare—only one other older than 3 million years (Lucy's) had been found at that time. Despite his caution, it became apparent in January 1995 that the team had found many parts of a skeleton. It was almost too good to be true. The Middle Awash team had found precisely the anatomy they needed to settle the question of whether *A. ramidus* walked upright: they had found the pelvis, leg bones (but not the top or bottom ends of a thighbone), foot bones, hand bones, and an ankle. They had also found the lower jaw with teeth, and a skull. White would exult that partial skeletons are the Rosetta stones of paleoanthropology. By seeing how bones link together, anatomists can reconstruct how those joints functioned—and re-create how their bodies moved, what the creatures ate, and what the relationship of their brain size was to their body size, among other questions.

The team's excitement was tempered, however, by the condition of the partial skeleton. It was the most fragile skeleton ever found. It would eventually include more than one hundred fragments so soft and crushed that they crumbled when touched. White speculated that the skeleton must have come apart while it laid on the surface, before it was buried. He soon realized that it made more sense to excavate entire blocks of sediment containing each of the fossils and to transport them back to the National Museum in Addis Ababa, where he could finish the excavation in the lab.

There, White used brushes, syringes, and dental tools to clean the encasing silty clay from the fragile bone, injecting it with a gluelike preservative so the bone would harden further as he worked on it. The skull was so squashed that the top of the cranium was only one inch above the bottom of the skull. And the bone was so chalky that he had to make a plaster mold of

Bryan Patterson found the first fossil of an early hominid at Kanapoi in the West Turkana desert in 1965. *(© William Sill)*

Louis and Mary Leakey holding the upper jawbone of *Zinjanthropus boisei* (renamed *Australopithecus boisei*), which they discovered in 1959. It was the first early member of the human family found in East Africa. *(The Leakey Foundation)*

Tim White (far left) meets for the first time with Donald Johanson (far right) in January 1976 in a museum laboratory in Nairobi. Also present are (left to right) Richard Leakey and Bernard Wood. Johanson had just returned from Hadar, Ethiopia, with new fossils of seven to twelve individuals, which were later determined to be members of Lucy's species. *(© David Brill)*

Kanapoi is a desert today, but woods once lined the shores of an ancient lake that filled the valley at about the time when upright-walking human ancestors lived there, 4 million years ago.

(© Ann Gibbons)

Meave Leakey returned to Kanapoi in October 2003 with Kenyan graduate student Fredrick Kyalo Manthi to collect fossils and samples of rocks to help reconstruct the ancient environment in which early human ancestors lived. *(© Ann Gibbons)*

Meave Leakey examines new teeth of the early human ancestor *Australopithecus anamensis*, which National Museums of Kenya associates Robert Moru and Justus Edung found amid the rubble at Kanapoi on this morning in October 2003. *(© Ann Gibbons)*

Meave Leakey painstakingly brushes away sediment to search for more fossils where the 4.1-million-year-old teeth of *Australopithecus anamensis* were discovered.

(© Ann Gibbons)

Tim White working in the Middle Awash of Ethiopia in 1996. *(© David Brill)*

(ABOVE, RIGHT) A 4.4-million-year-old child's jaw of *Ardipithecus ramidus* has a milk molar that is far more primitive than the tooth of any other early human ancestor. *(© David Brill)*
(BELOW) Yohannes Haile-Selassie and his colleagues collected these fossils of the early human ancestor *Ardipithecus kadabba*. It lived between 5.2 million and 5.8 million years ago in Ethiopia. *(© David Brill)*

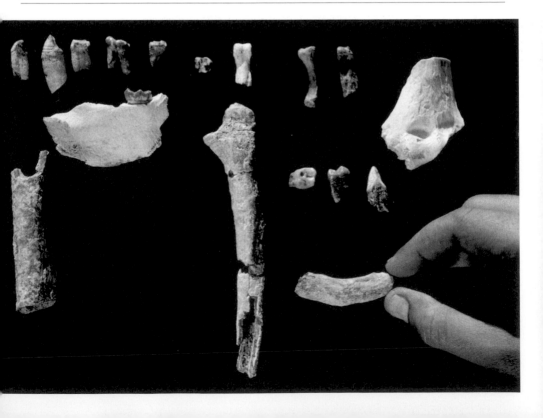

Yohannes Haile-Selassie (center) analyzes animal fossils found with *Ardipithecus kadabba* in the lab at the National Museum of Ethiopia, along with Scott Simpson (right), Cesur Pehlevan (left), and Leslea Hlusko (back), all members of the Middle Awash Research Group. The animal bones will help them reconstruct the type of forest where *Ardipithecus* lived. *(© David Brill)*

Yohannes Haile-Selassie and Giday WoldeGabriel explore the remote western margin of the Middle Awash in Ethiopia where they found fossils of *Ardipithecus kadabba*. This fossil site was much more elevated and lush when *Ardipithecus* lived here 5.6 million years ago. *(© David Brill)*

(LEFT) Brigitte Senut and Martin Pickford point out the spot where their team found the first fossils of 6-million-year-old Millennium Ancestor (*Orrorin tugenensis*) in the Tugen Hills, Kenya. *(© Ann Gibbons)*

(BELOW) Kiptalam Cheboi discovered the first fossil of *Orrorin tugenensis* in 2000 and a 4.5-million-year-old jawbone of *Ardipithecus* in the Tugen Hills in 1984. *(© Ann Gibbons)*

(BELOW) The owner of this thighbone (at left)—*Orrorin tugenensis,* also known as Millennium Ancestor—walked upright 6 million years ago. But does the bone show whether he walked like a modern human or in a more primitive manner?

(© Marc Deville / Gamma)

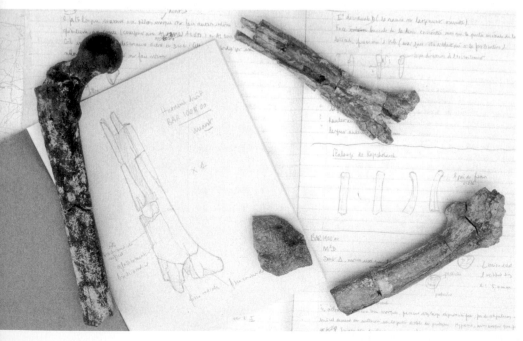

The Tugen Hills are a rare spot in Kenya's Great Rift Valley where a block of land was pushed up, exposing layers and layers of ancient sediments representing 16 million years of geologic history. (© Ann Gibbons)

(BELOW) Andrew Hill has led the Baringo Paleontological Research Project in the Tugen Hills since 1985; there, he and his colleagues have found a 4.5-million-year-old jawbone of *Ardipithecus* and fossils of extinct apes.

(Phil Crabb)

(BELOW) At Toros-Menalla, in the Djurab Desert of Chad, Michel Brunet (lying on ground, left) and members of MPFT sweep and scour the sand dunes for fossils. (© MPFT)

French paleoanthropologist Yves Coppens's first major discovery was the skullcap of Yayo in Chad in 1961. Later, he was codiscoverer of fossils of Lucy's species in Ethiopia. *(© Ann Gibbons)*

(ABOVE, RIGHT) A clay sculpture reveals how Toumaï might have looked when he was alive 6 million to 7 million years ago. He lived in a lush tropical forest that might have been similar to the Okavango Delta of Botswana. *(© MPFT. Scientific adviser: Michel Brunet; sculptor: Elizabeth Daynes)*

(BELOW) Michel Brunet (right) and Chadian graduate student Likius Andossa compare the skull of Toumaï —*Sahelanthropus tchadensis* (in Brunet's hands)—with a chimpanzee skull (by his left hand). *(© MPFT)*

every piece before he tried to reconstruct them into a partial skeleton—painstaking work that was like assembling a complex puzzle of thousands of pieces when many key pieces were missing. Later, Gen Suwa used an industrial computerized tomography scanner (CT scanner) to make digital copies of each piece. They also reassembled a virtual partial skeleton in their computers as another way to reconstruct it accurately.

White had just received the delivery of the first set of bones, which were still in their plaster field jackets to protect them for transport, when Meave Leakey came to visit him in Addis Ababa in January 1995. She had casts of the shinbone, jawbone, and teeth from hominids that were alive at Kanapoi 4.1 million years ago. Both White and Leakey were eager to compare the fossils that they had discovered to see if they were the same species—or were closely related. Even Henry Gee had prodded them to get together, writing in his New Year's piece in *Nature* that comparisons of the fossils from Ethiopia and Kenya would be necessary to establish whether the fossils represented one species or more, and whether *A. ramidus* was a hominid.

White and Meave Leakey had already planned to meet to compare the corresponding teeth, jaw fragments, and other bones in the two sets of fossils. It quickly became clear, for example, that they were looking at two different types of hominids—and that *A. ramidus* was more primitive than the fossils from Kanapoi.

Leakey and Walker would soon make an announcement about their team's discoveries at Kanapoi and Allia Bay, a bone bed on the eastern side of Lake Turkana where fragments of teeth and bones would be dated to 3.9 million years old. Later that year, in August 1995, they would describe twenty-two new fossils found at both sites that proved that they had discovered a new type of hominid that was older than Lucy's species. They named this new hominid *Australopithecus anamensis*. A shinbone showed that these ape-men or ape-women of the lake had walked upright 3.9 million to 4.1 million years ago. This would be the oldest unambiguous evidence for upright walking in a member of the human family, if indeed the shinbone was as old as they thought. Leakey and Walker also suggested that *A. anamensis* was the direct ancestor of Lucy's species, *A. afarensis*.

Eventually, Leakey would travel with casts of the fossils from Kanapoi and Allia Bay to Tempe, Arizona, where she would visit Don Johanson, the discoverer of Lucy. In the mid-1990s, after a contentious split with Berkeley geochronologists White and Asfaw, he had moved the nonprofit Institute of Human Origins, which he'd founded in Berkeley, to Arizona State University. There, Johanson and his longtime associate paleoanthropologist Bill Kimbel spread out the casts of fossils of *A. afarensis* on a long conference table, comparing them piece by piece with Meave's casts. By the end of the session, they agreed they were seeing one long-lasting lineage of early human ancestors that had evolved over time. Such a lineage can include several sequential species, which are known as chronospecies.

Soon after, White and his colleagues made a move that suggested that *A. ramidus* was in a class by itself. They had changed their classification at the urging of their colleagues and decided to move the fossils out of *Australopithecus,* as Gee had predicted, and into a new genus of its own—one they called *Ardipithecus,* drawing from the Afar word for "ground"; hence, it was now the root ground ape. The new name—*Ardipithecus ramidus*—would be published in *Nature* later that year.

Where there had been no hominid older than 4 million years, now there were two different genera. Some paleoanthropologists were beginning to wonder if they were seeing fossils that were part of a radiation, or a proliferation of different types of early hominids that walked upright in different ways—and if more than one had been alive at the same time 4 million years ago. This talk of a "bushy" family tree would drive White up a tree: he would point out that two species older than 4 million years was hardly a radiation, and that a radiation in the fossil record would look more like the famous example of the Burgess shale of Canada, where dozens of different species of strange spiny creatures lived together half a million years ago in an inland sea. Nonetheless, consensus about the neat line of descent from *A. ramidus* down to Lucy and eventually to modern humans was beginning to erode. Gee would predict, "The next few years could see violent upsets in our understanding of human origins."

CHAPTER ELEVEN

———————————

WEST SIDE STORY

The time had come for him to set out on his journey westward.

JAMES JOYCE

Yves Coppens was sitting at a table in an outdoor café in Paris, on the Île Saint-Louis, with a stunning view of the Cathedral of Notre Dame. It was a crisp and clear autumn morning in October 2002, and Coppens had walked to the café. He was carrying a conspicuous white paper bag, which he placed carefully beside him on a chair. He ordered only water and began to talk about his first trip to Chad, in 1960, when he was a young man of twenty-six and Chad had just been granted independence from France. As he reminisced about exploring the Djurab Desert in an ancient Dodge Power Wagon, with a crew of four and a two-way radio that often was out of range of help, he would occasionally pat the bag on the chair beside him, as if to make sure it was still there. He was one of the first paleontologists to search for fossils in the Djurab Desert, where he had discovered the sandy fossil beds that his longtime friend Michel Brunet would explore three decades later. In seven expeditions to Chad, Coppens discovered thousands of fossils. As he described the extinct animals he had found, he casually reached inside the shopping bag and pulled something out. He held it up, so that it looked like an unimpressive lump of rock. Then, like a magician, he turned it over: voilà, there was a face that looked almost human. Eye sockets filled with sandstone stared across the table. "This is Yayo," he said, with a twinkle in his eye. It was a trick he had done before, and he clearly still enjoyed its effect on the uninitiated.

As he held it up, other Parisians sitting at nearby tables began to notice Coppens and the skull he had pulled out of his bag. At sixty-eight, Coppens was white-maned, with droopy eyelids that often look bemused, and a contained manner that gave him an almost courtly presence. One or two people in the café seemed to realize that he was someone worth recognizing, even though they could not quite place him. Indeed, he has been the most famous paleontologist in France for twenty-five years, with as many medals as a war hero—he is a knight of the Legion of Honor, an officer of the National Order of Chad, the recipient of the gold medal of the emperor of Ethiopia, and the beneficiary of two dozen prizes and honors from museums, universities, and corporations in Europe and America. He has published books on human evolution and been a consultant on everything from prehistoric novels to television documentaries. He became France's leading spokesman on human evolution when he was still in his thirties, soon after he had discovered some of the first fossils of Lucy's species as part of the international team that worked at Hadar in Ethiopia.

But his first big discovery was Yayo. His wife, Françoise Coppens, discovered it in 1961 and named it *Tchadanthropus uxoris,* which means "my wife's Chad man." It was the top part of a skull and face, and the first hominid found in Chad, only two years after Louis and Mary Leakey had found their first hominid fossil—Zinj—in eastern Africa. They found Yayo upside down—as he demonstrated in the café—and did not see its importance until Coppens turned it over and was face-to-face with an early human. Today, he thinks it is a specimen of *Homo erectus,* a direct ancestor of humans, and it was probably living on the shores of the ancient Lake Chad about 1 million years ago. It would be the only hominid found in Chad for three decades.

Coppens worked in Chad until 1967, when it became too perilous. Indeed, soon after he left, a French archaeologist, Françoise Treinen-Claustre, was kidnapped by northern tribes and held captive for three years. Civil war broke out between the northern Muslim tribes and the government, which was dominated by the non-Muslim Sara people in the south. Chad soon fell into a pattern of military crackdowns and attempted coups—turmoil that was aggravated by conflict and occupation of parts of the nation by Libya, its aggressive northern neighbor. Soon, this former French colony would become one of the poorest nations in the world, ravaged by conflict and recurring droughts that persisted for three decades. By necessity, Coppens shifted his focus to Ethiopia, where the politics under Emperor Haile-Selassie were comparatively calm. So, in 1967, he began work as coleader of the Omo Expedition, and he later joined the expeditions to Hadar that found Lucy and her species.

While working in the Omo Valley, Coppens would have plenty of time to reflect on the small number of hominid fossils he and his colleagues were finding compared with the tons of bones of animals, both in Chad and Ethiopia. He would do the math and calculate that for every hominid fossil found at Omo, for example, paleontologists would find ten thousand fossils of extinct animals—a ratio that underscored how vanishingly scarce human ancestors were on the landscape 3 million years ago. Their populations had obviously been small, and hominid bones had never "littered" the landscape.

By the early 1980s, researchers had begun to publish their inventories of fossils collected during the rush of expeditions to Africa in the 1960s and 1970s. They had collected two thousand fossils of hominids in eastern Africa in twenty years, ranging from isolated teeth to partial skulls, and hundreds of thousands of fossils of animals. But as rare as hominids were, there was not a single fossil of a chimpanzee or a gorilla in eastern Africa.

Coppens was thinking about this puzzle during a conference on human evolution held in the Papal Academy of Sciences at the Vatican in May 1982 when he opened an atlas showing the present distribution of animals in Africa. It reminded him that chimpanzees and gorillas live in the large forests of tropical Africa, but not in the more open terrain of the Great Rift Valley to the east. But he knew that all the fossils of hominids that were 3 million years or older had come, without exception, from the east side of the Great Rift Valley. The skull he had found in Chad, to the west, was 1 million years old at most.

As he looked at the atlas, he began to formulate a model that would explain the geographic divide between the early ancestors of humans and African apes. He would propose this hypothesis three years later, in a lecture at the American Museum of Natural History in New York City. He began with a scenario 8 million to 10 million years ago in the tropical forests that covered equatorial Africa, from the Atlantic to the Indian Ocean. This was the home range of the common ancestor of humans and chimpanzees until the movement of the tectonic plates and the erupting volcanoes remodeled the landscape and created the Great Rift Valley, which bisects eastern Africa. While most paleoanthropologists had focused on the eastern rift valley within the Great Rift, geologists had produced new dates for the formation of another rift valley to the west—the western rift, which began forming about 8 million years ago. It is one of the great dividing lines of Africa, and can easily be identified on a map because its low valleys include the basins for a chain of African Great Lakes, including Tanganyika, Kivu, Rutanzige, and Albert. The western rift was created when shoulders of land bordering its western rim were pushed up into tall mountains, including the mist-enshrouded Ruwenzori Mountains along the western border of Uganda. The combination of the high mountain barrier and the low valley floor disturbed the circulation of air, trapping rain clouds on the western flank of the western rift. The climate changed, with land on the western side of the rift remaining wet and humid, while the land to the east became hot and dry. The ancestors of apes, such as chimpanzees, stayed in the forests west of the rift zone. All of this was known from the geology.

Coppens added this new wrinkle: he proposed that the ancestors of humans began to adapt to their new life in a drier, more open environment. They ventured east into the rift, and those that could adapt to a broader range of habitats were more likely to survive, reinforcing that adaptability in later generations of hominids. This compelling model would explain simply and clearly why chimpanzees and humans were so close genetically, but why the fossils of their ancestors were never found in the same terrain. Coppens called it the East Side Story of human evolution. It was a hypothesis that begged to be tested. All it would take to prove the East Side Story wrong would be to find a fossil of an African ape on the east side of the Great Rift Valley—or a fossil of an early hominid on the west side.

☒

Michel Brunet would be the man to test Coppens's hypothesis. Brunet regarded Coppens as a sort of professional older brother—a successful scientist six years his senior whose advice he sought and appreciated, but whom he hoped to outperform as well. Coppens, for his part, had been mentor to Brunet and an entire generation of paleontologists and paleoanthropologists in France, and had been generous with his letters of recommendation to work in foreign countries, to obtain funding, and to get promotions. Brunet would turn to Coppens more than once for advice, particularly when he decided to make a radical change in his career in the mid-1970s—to join the quest for early human ancestors. It was a momentous decision for Brunet, and he has taped an old permit allowing him to work in Afghanistan on his office door at the University of Poitiers. It is posted at eye level, where it is hard to miss the photo of Brunet at the age of thirty-six, with dark hair and a beard, eyes smiling. The permit was issued in 1976, and Brunet sees it as a sentimental reminder of a critical turning point in his life. "This is when I decided to change my topic," said Brunet, as he pointed to the permit and the photo of himself. "I was a young guy."

In 1976, Brunet was a vertebrate paleontologist who studied artiodactyls, the family of hoofed ruminants that includes cows, pigs, sheep, antelope, giraffes, and many of the large land animals. He had spent years tracking the migrations of these mammals into Europe 40 million years ago, and had published papers with titles such as "The First Discovery in Europe of the Skull of an Entelodont" and "A New Primitive Ruminant from the Oligocene of l'Agenais."

It was a respectable career for a local boy who had spent his early childhood in the small village of Magné, forty kilometers south of Poitiers, in the heart of old France, what the French (and Brunet) refer to as *La France profonde*. He was born in 1940, only a few months before the German army swept into Paris. His parents, who lived in Versailles near Paris, soon sent him to live in the country with his maternal grandmother, because they thought it would be safer there during the German occupation. He spent his early years outside, as a young shepherd boy, exploring the countryside with his dog, observing horses, cows, and other farm animals. It instilled in him a deep love of nature and animals, and for a while he aspired to be a veterinarian. "I am happiest when I sleep under a blanket looking at the stars," he says. He did not start school until he was eight years old, when his parents brought him back to live with them in Versailles.

Despite his late start in school, he did well in math and science and entered the University of Paris, earning a Ph.D. in paleontology. He would say years later that a lecture by Louis Leakey had influenced him more than any other talk or paper he heard in his career. He heard Leakey give a paper in 1963 at the National Museum of Natural History, in which he introduced the new species *Homo habilis.* That lecture, as well as the books of Charles Darwin, inspired Brunet to study human evolution. But like many French paleoanthropologists of his generation—including Coppens—Brunet started his career by working with animal fossils first. In France, the tradition has been that paleontology students examine animal fossils first, because they are far more numerous than hominid fossils and better specimens for learning how to work on fossils without damaging them. Their sheer numbers also make it easier to recognize similarities and differences in anatomy within one species, for example, whereas in hominids some species are represented by only a handful of teeth and jaw fragments. After college, Brunet returned to Poitiers, becoming a professor of paleontology at the University of Poitiers, because he missed the quality of life in a small town. He married, had two children, and settled into life in the country near Poitiers.

By the time he was thirty-six, he was restless. "What does he want?" Brunet asks rhetorically, speaking of himself. "I am *Homo sapiens,* and I want to try to answer: From where are we coming?" He would not be able to answer that question studying mammal migrations in Europe. At the time, one of the frontiers of exploration for early hominids was in the Siwaliks of Pakistan, where David Pilbeam's team had found a new lower jaw of *Ramapithecus,* which was still considered by many to be the earliest known member of the human family. Brunet made a momentous decision: to seek extinct apes himself, in the sediments across the border from Pakistan, in Afghanistan.

Finding hominids is not as easy as collecting shells at the seashore, as Brunet is fond of saying. He and his colleague Emile Heintz, a paleontologist at the Centre National de la Recherche Scientifique (CNRS) in Paris, worked in Afghanistan in 1977, 1978, and 1979. They found rodents and an extinct baboon, but no apes. Brunet was in his element, however, enjoying life as a nomad, getting to know the Afghan people. But it was risky business: in 1978, Brunet and his colleagues were standing on a flat rooftop in Kabul, watching Russian fighter jets bomb the city. One jet roared low overhead, three times. The fourth time, the plane fired on them. Fortunately, the paleontologists had run for cover and the pilot missed.

Brunet also started fieldwork in Iraq. That ended quickly one night when he was arrested while seeking a hotel room in a small town. He never found out the reason for his arrest. After he was released, he could not return to Iraq. It also had become clear to Brunet and Heintz that they would not find hominids in Afghanistan, because the fossils of mammals they had discovered were not the type usually associated with early hominids. Brunet had a new idea: he thought it would be interesting to compare the fossil animals found in Afghanistan with those that were coming from the fossil beds of the same age in Pakistan where *Ramapithecus* had been found. He asked Coppens to write him a letter of introduction to Pilbeam, seeking permission to join the team in the Siwaliks. Pilbeam said yes. Brunet and Heintz soon found themselves at the Silver Grill Hotel in Rawalpindi, an outpost near the Siwaliks, where they met Pilbeam for the first time. It was the beginning of a long friendship.

Brunet worked in the Siwaliks in Pakistan in the early 1980s, just as Pilbeam and his colleagues came to the conclusion that the new fossils of the face of *Ramapithecus* showed that it was the extinct ape *Sivapithecus*, not a hominid. This meant that there was no evidence for hominids outside of Africa before 2 million years. It was clear that the search for the earliest ancestors of humans and their closest cousins—the African apes—would have to shift back to Africa. But where should they focus in Africa? Most of Africa was totally unknown to paleoanthropologists. The researchers who had worked in the Siwaliks would continue to focus on slices of time from the Miocene, about 5.3 million to 23.8 million years ago, but in different places in eastern Africa. Pilbeam and Andrew Hill at Yale would start the Baringo basin project in Kenya; Martin Pickford would go to Uganda. Pilbeam would also propose the most radical idea—he would search for the ancestor of African apes in the forests of western Africa. "I'd already figured out it would be important to find fossil chimpanzees," Pilbeam would recall. "How would you recognize an early hominid if you didn't have a chimpanzee ancestor for comparison?"

Pilbeam studied ecological maps, searching for places where there might have been ancient rain forests. His strategy was to look for the rainy side of volcanoes, where rainwater would have collected to support rain forests—and where there would be volcanic materials to date the sediments. He lighted on Cameroon, which was on the southwestern border of Chad. It was a nation with close ties to France, with its currency tied to the French franc, making it a logical place for French researchers to get funding and logistical help. Even the food and champagne in the part of Cameroon that had been French before independence in 1961 were better than that in the part of the nation that had been British. Pilbeam immediately thought of Brunet, and asked him in the summer of 1981 if he wanted to search for fossils in the formerly French part. Brunet responded enthusiastically, and eventually the two got a field permit to go there in 1984. Many of their colleagues thought they were crazy, because no one ever finds fossils in forests. "We were two young guys. One British, one French. We were crazy," Brunet would recall.

They did find sites of the right age. But the work in Cameroon was discouraging. They soon realized that their colleagues were right—there were practically no fossilized bones. The soil was too acidic, and almost all of the fossils had disintegrated long ago. In nine expeditions between 1984 and 1993, Brunet and his colleagues found fossilized leaves and insects, but only a single fossil from a mammal. Their biggest discovery was dinosaur footprints. It was ironic: Meave Leakey had spent years working at a place— Lothagam—where the fossils were well preserved but where there were probably no hominids. Brunet spent years searching for hominids in precisely the type of forests where apes—and perhaps the earliest hominids—were more likely to have evolved, but where all traces of their bones had vanished. Either way, no fossils of hominids meant no funds. Pilbeam dropped out. Brunet stayed, but he paid a high price. In 1989, a close friend of his, geologist Abel Brillanceau, died of drug-resistant malaria when he was with Brunet in Cameroon. It was devastating for Brunet, who more than a decade later would still feel deeply responsible for the death of a member of his team. Brunet would note that Brillanceau was the father of five young children, and that their loss still haunts him. He would forevermore be aware of what could go wrong when he invited colleagues to join him to work in some of the most dangerous places on the planet. The allure of Cameroon would wane. Pilbeam would say: "We both decided Cameroon had passed its sell-by date."

On July 14, 1994, Brunet and Pilbeam were on a friend's balcony in Paris, watching fireworks. It was Bastille Day, and Brunet had a revolutionary plan of his own. He turned to Pilbeam and said: "Come on. You have to come to Chad!" Although there had been at least two attempted coups in Chad in 1992 and 1993, the country was in a period of relative stability under the pragmatic leadership of military adviser Idriss Déby, who had been president since 1990. Brunet had already gone on one exploratory mission in January 1994, and had found fossils of animals that were the same age as Lucy or slightly older—between 3 million and 4 million years old. But for Pilbeam, Brunet might as well have asked him to storm the Bastille. He had many responsibilities as dean of undergraduate education at Harvard University, and he didn't want to be away from his young daughter and wife, a molecular anthropologist whose case for tenure at Harvard was pending. This time, he said no.

Brunet persevered. During the last years he had worked in Cameroon and in Nigeria to the northwest, he had explored the terrain near their borders with Chad, on the shores of Lake Chad. He was eager to push north into Chad, to survey the fossils along the old lakebed of the much larger Mega-Lake Chad, which had existed millions of years ago. But constant warfare made it too dangerous. Then, in 1992, he got the break he had been waiting for: a French geographer working in Chad invited him to give a lecture in N'Djamena, the capital of Chad. The geographer, Alain Beauvilain of the University of Paris X–Nanterre, had been posted by the French foreign service to Chad to help build the research infrastructure of the nation in 1989. He was asked by the Chadian government to create the National Center for the Promotion of Research, and as director of that center he was encouraging foreign researchers to do research in Chad, including geologists who searched for meteorites and volcanic craters. While Brunet was in N'Djamena, Beauvilain told Brunet it would be possible for him to go into the desert to search for fossils—if he could get his own funding. Beauvilain would help him get permission from the Chadian authorities and would help organize the expedition.

It took Brunet two years to scrape together funds, but in January 1994, Brunet and Beauvilain set out with a small crew of geologists, a botanist, and a guide in three four-wheel-drive vehicles to explore the Djurab Desert near Koro Toro, a site that Coppens had found decades earlier in the center of Chad. The French government paid Beauvilain's expenses and provided a vehicle. The team of eight also borrowed a vehicle from the Chadian Ministry of Mines and Energy. Brunet had appealed for funding wherever he could, even convincing local politicians from his town of Poitiers to give him a grant to study the effects of goat grazing in Chad. His tenacity paid off: on the first trip, the survey team found eleven fossil sites east of Koro Toro where the animals were 3 million to 4 million years old.

Brunet returned the next January without Pilbeam, but with Beauvilain and an even smaller team. This time, their intent was to push much farther north, near the cliffs of Angamma, which in the local Goran language means "outside tombs." This aptly named site was where Coppens had found his partial hominid skull and upper face three decades earlier. To their delight, they found an even better site on their return trip in the region of Bahr el Ghazal (Arabic for the "river of the gazelles"). On January 23, 1995, they discovered the partial jaw of a much older australopithecine that was alive 3 million to 3.5 million years ago—the one Brunet named Abel, for his former colleague who had died in Cameroon. He would announce the discovery of an australopithecine twenty-five hundred kilometers west of the Great Rift Valley in November 1995 in the journal *Nature*. Later, in May 1996, he would name the jawbone *Australopithecus bahrelghazali,* describing it as a western cousin of Lucy's—two separate but closely related species of early humans (although many paleoanthropologists thought it belonged in Lucy's species).

〰

As news went around the world of the discovery of an early hominid in Chad, paleoanthropologists outside of the small community of paleontologists who study the origins of mammals were asking, Who is Brunet? It was as if he had come out of nowhere, with the first early hominid outside of eastern Africa and South Africa. "The two windows that were open to us—South Africa and eastern Africa—are only a very small part of the picture," observed paleoanthropologist Don Johanson, who discovered Lucy. "Brunet had opened a third window."

The single jawbone showed that hominids had ranged from eastern to western Africa for at least 3.5 million years. It didn't disprove Coppens's East Side Story, because this new western australopithecine was about a million years younger than the oldest known hominid from the east side, the 4.4-million-year-old *Ardipithecus ramidus* discovered in Ethiopia by the Middle Awash team. "We're not saying we know where the cradle of humanity is, but the cradle is much larger than we thought," Brunet told scientists at a press conference in Paris in May 1996 when he named the new species. Brunet had seen enough of the fossil beds of Chad to know that he had barely begun to scrape the surface of the dry lakebed of the ancient Mega-Lake Chad.

The discovery of Abel would also pave the way for Brunet to win backing for future expeditions from the CNRS; the French ministries of education and foreign affairs; the regional government of Poitou-Charentes; and the city government in Poitiers—he was even awarded the Philip Morris Scientific Prize. In the two years after discovering Abel, the Mission Paléoanthropologique Franco-Tchadienne (MPFT), as his loose-knit team of French and Chadian researchers was now called, returned to the field to search for fossils more than a half dozen times. Brunet bought a lightweight plane for airborne reconnaissance, although it would prove too delicate for the harsh conditions of the Djurab and was often grounded. They also abandoned car camping except on lean exploratory missions, in favor of expedition-style tents—including a base camp erected one year by the French army, whose soldiers put up four long tents, unloaded pallets of Evian water, and scoured the dunes for fossils. But even with such organized support, the logistics were daunting: two tents were destroyed by sandstorms that persisted without respite for three weeks. The tents had been installed at an angle that turned them into windbreaks, which meant that they became the base for the formation of new sand dunes—and team members had to dig out of the sand as if they were snowed in. They learned that year to set up tents in the protected hollows between dunes.

Even as Brunet's expeditions became more sophisticated, he was forming strategic alliances in the scientific community. He brought colleagues and graduate students from France, the University of N'Djamena, and Harvard to the field in Chad. In 1997, Pilbeam visited for a week. "The fraternity is more important than everything else," Brunet would say. "Without my team, I am nothing."

He also traveled to Nairobi to compare Abel with the fossils that Meave Leakey and Alan Walker had found near Lake Turkana, forging a friendship with Meave and vacationing with the Leakeys on the Kenyan coast. He met with Berkeley paleoanthropologist Tim White and his team members in Addis Ababa in July 1995. At the same time, he noticed how White and Berhane Asfaw had set up the NSF lab in Addis Ababa, and how the Middle Awash team was recruiting and training Ethiopians. He was a quick study and returned to Chad with funds to build a museum lab in N'Djamena, and to fund scholarships to train Chadian students in Poitiers. And he arranged to send some of his students to Berkeley and Harvard for postdoctoral fellowships, to make sure that his team was on the cutting edge of analyzing the fossils of animals, as well as of hominids.

By early 1997, he knew that the Franco-Chadian team was closing in on the right time and place to find out if hominids were in Chad as early as they had been in eastern Africa. The team had revisited an area called Toros-

Menalla in January 1997 that Coppens had found, and they were discovering fossils of animals 6 million to 7 million years old. He had a feeling that they were on the verge of something important: "Sometimes in the field you see something, and you have a feeling it is a good place," Brunet would say later. He was so confident that even his team members were convinced they would find the earliest hominid. But just as they were beginning to survey these older fossil beds, Brunet had a heart attack. Years of smoking and pushing himself to the limit, both at home and abroad, had taken their toll. Fortunately for him, the heart attack occurred in Poitiers, rather than in Chad. He ended up having emergency bypass surgery; his arteries are now kept open with the assistance of several stents.

Brunet was unable to return to Chad for twenty-two months, although the team continued to press on without him—the Mission Paléoanthropologique Franco-Tchadienne had taken on a life of its own. But from that point on, Brunet would express a sense of having too little time. Success had come to him relatively late in life. It was bittersweet, because he knew he would never live long enough to achieve all that he wanted to do in Chad.

CHAPTER TWELVE

_____

TURF WARS

Under the unspoken rules of the paleoanthropological game,
one does not move in uninvited and begin picking nuggets out of
another prospector's claim.

DONALD JOHANSON, 1981

In 1995, Martin Pickford launched a surprise attack on Richard
Leakey. He published a book with a Kenyan colleague that was called *Richard
E. Leakey: Master of Deceit*. The cover showed a drawing of Richard Leakey
holding a skull with green dollar bills falling out of it. The title and cover
were tame compared with the contents. The book read like pulp fiction, and
even described in embarrassing detail the alleged sexual indiscretions of several researchers. It started with a dedication to "the victims of the fantastic
Richard Leakey manipulations." It described Leakey as a "parasite," his science as "folly," his books as a "swarm of errors," and his friends as "toadies."
And that was just in the first two pages. In 218 pages of tightly packed text,
the book relentlessly attacked Leakey and his close colleagues, accusing
Leakey of all sorts of schemes to acquire fame and fortune. Some of the stories were punctuated with photocopies of Leakey's correspondence with dis-
gruntled museum employees and foreign researchers, including a notably
heated exchange with Berkeley paleoanthropologist Tim White.

The self-published book might have slipped under the radar of many researchers, except that Pickford and his Kenyan coauthor, Eustace Gitonga, orchestrated a campaign to introduce the book on the BBC news in London, then to have it serialized in the *Times* of Kenya. As word of the book spread, most paleoanthropologists thought that Pickford had lost his senses. Even those who were sympathetic to Pickford's plight thought the highly personal attack on Leakey and other prominent paleoanthropologists was a mistake—one that would marginalize Pickford even more, and that would further tarnish the image of the field of paleoanthropology. Pickford's undergraduate adviser, Canadian paleontologist Basil Cooke, who had had his own well-known disagreement with Leakey on the dating of a site in Kenya, would send Pickford a one-line letter that said: "I do not approve." Coppens, who had helped Pickford get a position at the Collège de France, would refuse to read it. And they were Pickford's mentors.

Richard Leakey would not dignify it with a response. By then, Leakey had entered the equally ruthless world of Kenyan politics to form an opposition party, so he was used to battling formidable foes. Tall and powerfully built, even after losing both legs below the knee in the 1993 plane crash, Leakey seemed to take pressure in stride. Pickford's book was just a taste of what was to come—he would take on the dangerous task of heading the government's anticorruption campaign, persevering even after getting beaten up by thugs. But though Leakey could take the heat, many of his close associates who were roasted in the book felt burned. Meave Leakey would dismiss Pickford as "evil." Andrew Hill would say, "Pickford truly enjoys screwing people." British paleontologist Alan Walker still fumes that the book said he had married his American wife—the well-known anthropologist and author Pat Shipman—so he could get grants from the U.S. government. (Walker has held chairs at two universities and can get U.S. grants without being married to an American.)

Pickford would say years later that he was not proud of the book. But he would also say, "I only have one regret—that it was necessary to write the book. If Richard had been a reasonable man, I wouldn't have had to write one." By the time he undertook the book, he had already been banned from working in Kenya because of the charges that he had tried to steal notebooks from the National Museums of Kenya. In that time, he had written numerous letters, protesting his innocence and saying that he had been denied due process. But he got no answer from the National Museums' trustees. Only Leakey responded, and he made it clear that the museum was still off-limits to Pickford—even after Leakey was no longer director. (Leakey resigned in 1989 to run Kenya's Wildlife Service.)

At the same time Pickford was getting nowhere with the National Museums, his rivals were finding fossils in sediments that Pickford was eager to explore. He had watched the earliest hominids creep back in age almost to the Miocene, which was his specialty. He was eager to get back to the Tugen Hills to see what more he could find there at the same time that *Ardipithecus ramidus* was alive in Ethiopia, and earlier. And he felt that he had been unfairly maneuvered out of his claim to the area when Pilbeam and Hill had launched the Baringo Paleontological Research Project.

By the time Pickford was approached by Gitonga to help him write the book about Leakey, Pickford said he was "fed up" and felt he had little to lose by becoming the coauthor. Gitonga is an artist who was the former head of the exhibits department at the National Museums of Kenya. Leakey had fired Gitonga from the museums in 1987 after accusations that he had misused funds to erect a large sculpture of a dinosaur in front of the museum. Gitonga denies the charges. He wrote most of the book attacking Leakey, and asked Pickford to edit it and help him get it published. Not only did Pickford add his name and a few chapters to Gitonga's book, but he also went on to write a second book, this time criticizing Richard's father, Louis Leakey (entitled *Louis S. B. Leakey: Beyond the Evidence*).

The books were just the beginning of their tireless campaign to circumvent Leakey and to end the National Museums of Kenya's monopoly on paleontology research in Kenya. Pickford later described the strategy in his journal, comparing Richard Leakey to a giant boulder in the road. The best way forward, he said, would be to go around it, leaving it to one side of the road.

⸺

As Eustace Gitonga showed a visitor his office on the twentieth floor of a high-rise building atop Nairobi Hill in September 2003, he proudly pointed out the sweeping views to the southwest of the Ngong Hills. "You can see Mount Kilimanjaro on a clear day," he said. This is the headquarters of the Community Museums of Kenya, and from his seat atop Nairobi, Gitonga and his colleagues outlined their vision to "bring museums to the people." Gitonga is a small man, neatly groomed, deliberate, with a close-cut haircut, immaculate clothes, and a wary smile. He is proud of his accomplishments—the opening of two museums in one year. One was the Kipsaraman Museum, high atop a ridge in the Tugen Hills, a large, single-room building that showcases fossils of hominids and animals found in the Baringo basin. The other was the Community Museum and Reptile Park built nearby, on the shores of Lake Baringo. He is aware that his critics don't take his plans for Community Museums seriously—that they consider the organization a front for an operation whose real purpose is to give permits to Gitonga's friends and to make life difficult for Leakey and his "cartel," as Gitonga calls Leakey's longtime colleagues.

But Gitonga does have a grand vision for Kenya. He plans to build twenty-three museums in Kenya, and as he talks, he displays drawings for the next one on his list—a museum for the Masai, south of Nairobi. To the north, he wants to erect an astronomical observatory on the slopes of Mount Kenya where Kenyans can be trained to become astronomers. The idea is to use the museums to train Kenyans to become Ph.D. researchers who eventually will take over management of the museums and research.

Gitonga says he came up with the idea of forming Community Museums of Kenya while he was traveling in Egypt soon after the publication of *Master of Deceit*. He says that while he was there, he noticed that there was a museum in almost every town or village, celebrating the heritage of the Egyptian people and bringing tourists to far-flung places where they could see Egyptian relics in the same places where they were discovered. Gitonga realized that Kenyans could also build community museums where they could display their relics—*including fossils*—locally, instead of sending them to Nairobi to the National Museums of Kenya. Under Kenya's antiquities laws, the National Museums of Kenya had been given the authority in 1983 to recommend approval or rejection of all paleontology research permits before they could be issued by a government ministry.

According to those laws, all fossils discovered are the property of the government, and the government institution responsible for them has been the National Museums. What Gitonga and his colleagues also realized soon after his Egypt trip was that by creating a rival system of museums with the authority to issue its own research permits, Community Museums could circumvent the National Museums—and Richard Leakey.

When he returned to Kenya, Gitonga quietly proposed his idea to politicians he knew. Gitonga's pitch was that Community Museums would recapture the national heritage of Kenya from the National Museums, which had been "exploited" by Leakey, a native Kenyan, and a small group of foreigners who were his colleagues. Gitonga was well connected politically, and he soon got permission to register the Community Museums as a nongovernmental organization (NGO) with the government. He got politicians' attention with a statistic: by 1995, only two black Kenyans had earned Ph.D.s in anthropology, yet dozens of foreigners had used Kenya's fossils for their doctoral dissertations.

Even though Gitonga often expresses disdain for foreigners (mainly Americans) who he says exploit Kenya's culture and fossil heritage, he sought help from European and Japanese researchers to make his vision a reality—asking Senut and Yves Coppens, as well as a Spanish researcher, to serve on his eight-member board. Foreign investment was critical for funding Community Museums, and foreign scientists were essential for sponsoring and training Kenyan students. But as he described his plans in 2003, only one Kenyan student was being trained in anthropology by the Community Museums—by Pickford and Senut (although four others were training to be fossil technicians). Gitonga was beginning to recognize that it was not easy to train students in science, but it was a start. He wrote in a letter to the journal *Science* that "scientists of good faith from anywhere in the world now have a choice to carry out research in Kenya, something that was not possible for the first 35 years of the country's independence." He added, "No longer is paleontology in Kenya the monopoly of a single family or institution. Kenyans have recuperated their heritage, and with the help of the international community, will use it for development of its human and scientific potential."

One of his first acts as director of Community Museums was to apply for research permits for Pickford to work in the Baringo area; later, he applied for research permits for a Japanese team doing geology in Baringo. Gitonga also recruited a key ally to his board of directors, a politician named Andrew Kiptoon from the Baringo basin. Kiptoon was a research minister and parliamentary deputy from the Baringo North district, which includes the Tugen Hills fossil sites—and a member of the same tribe as former President Moi. In 1997, Kiptoon encouraged Pickford to begin developing the archaeological and paleontological potential of the Tugen Hills. In a letter of June 1998, Kiptoon wrote that it would not be difficult to obtain the necessary clearance, work permit, and other approvals. Indeed, three months later, Pickford got what he had been seeking for two decades: an official in the department for issuing research permits (then housed in President Moi's office) had approved Pickford's application to conduct paleontology research in three Kenyan provinces, including the Tugen Hills.

On January 4, 1999, Pickford met with Kiptoon, who gave him the "green light" to start development of the Tugen Hills, including plans to build the museum and roads and open other tourist attractions, such as a gulch with fossilized trees that could be billed as a petrified forest. Two days later, Pickford was in the Tugen Hills, meeting with the local chief of the Tugen people and laying the groundwork to search for fossils in several places, including Tabarin, where Andrew Hill and his colleagues had found the jawbone in 1984.

Less than two months later, Andrew Hill was at work in his Yale office when he got a telephone call from a friend who had picked up the French magazine *Le Point* in the Charles de Gaulle airport in Paris. The friend was Craig Feibel, a geologist at Rutgers University who had dated Meave Leakey's sites at Kanapoi and who was traveling home to the United States by way of Paris. He was surprised to read in the February 6, 1999, issue of the magazine that Martin Pickford said he was resuming work in the Tugen Hills; the article showed photos of him scouting a site for fossils there. "What's going on?" Feibel asked Hill. The phone call caught Hill by surprise. He had heard that

Pickford had been seen driving around the Tugen Hills, and Pickford had signed the guest book in the nearby branch of the National Museums of Kenya. But as far as Hill knew, his Yale team had the only permit approved by the government to do research and a second permit required by Kenyan law to search, excavate, and collect fossils in the Tugen Hills. He immediately contacted the National Museums.

When the National Museums officials heard about Pickford's permit, they challenged its legality, saying that Kenyan law required that Pickford also obtain a separate excavation permit to remove fossils. The director of the National Museums at the time, George Abungu, demanded that the research permit be revoked. Indeed, the same government official who had granted Pickford's permit on October 30, 1998, signed a letter revoking the permit.

But Pickford continued to work in the Tugen Hills, and no one bothered to stop him. He said later that he did not get the letter revoking his permit until November 1999, when a museum researcher handed it to him. Pickford immediately questioned whether the letter was legitimate, writing in his journal that day that "the letter is a crude and inept forgery." He saw more than one version of the letter, which had different comments in the margins. What made him suspicious was that the same government official who issued the actual permit with Pickford's photo, on November 30, 1998, had revoked it twenty-eight days *earlier*, on November 2, 1998. The official, Josephat Ekirapa, told *Science* in 2001 that the letter revoking the permit was not a forgery, but that he might have pre-dated the revocation notice after signing the actual permit—perhaps after the museum officials complained in 1999. In any case, Ekirapa said that if Pickford thought the letter was a forgery, he should have spoken with him—and that he had no excuses for continuing to work in the Tugen Hills after seeing the letter.

Pickford knew that the National Museums officials were challenging his permit by August 1999, because he wrote in his journal that the Tugen Hills politician Kiptoon had told him he had received a letter saying that Pickford's research permit had been revoked. Pickford met soon after with Kiptoon, and left with reassurances that Kiptoon "supports us to the hilt" on this, as did other local officials, he wrote in his journal.

Pickford also hired the local Tugen man who had worked as a fossil collector since 1968 with Andrew Hill—Kiptalam Cheboi, who had recently retired from the National Museums but was still employed by Hill. Kiptalam was good at spotting fossils; he had found the Tabarin jawbone when he was working for Hill. Pickford had known Kiptalam for many years, since his graduate school days when he started exploring in the Tugen Hills. Pickford, Senut, and Gitonga also repeatedly claimed that Hill had not worked in the Tugen Hills since 1993, implying that if he didn't search there enough, it would be fine for them to step in. "There are so many fossils, I don't know what Hill was doing all that time," said Pickford.

Such statements make Hill seethe. "I resent having to document how much time I spent in the field," said Hill, who had worked in the Tugen Hills every year except for one during the decade before Pickford got his permit. Nonetheless, to show that Pickford and Senut were wrong, he went ahead and documented his time in the field with evidence from his field logs and testimony from his team members. At the very least, Hill said, Kiptalam must have told Pickford the extent of his team's work in the area. Hill and other members of the Baringo Paleontological Research Project also had results to show for their time and work there—they published detailed descriptions of the fossils and geology of the Tugen Hills in a special double issue of the *Journal of Human Evolution* in 2002, noting that the project had been investigating the Tugen Hills continuously since 1981.

Despite the strength of Hill's claim in the Tugen Hills, Pickford's bold takeover bid succeeded: government officials could not agree on whether Pickford had valid permits or not and whether he needed the second excavation permit. In the absence of agreement—or enforcement of those laws—Pickford continued to make inroads in the area. His team hired local people to build new roads to fossil sites and to construct a museum, as well as helping them search for fossils. Pickford and Senut would become entrenched in the Tugen Hills.

It would not be enough for Pickford to work in the Tugen Hills. With his permits, Pickford was able to test the limits of his new Community Museums permit, which gave him access to do research in the Rift Valley province of Kenya as well. He went to Kanapoi and, in effect, had a head-on confrontation with the Leakeys.

By early 2000, Meave Leakey, Alan Walker, and their colleagues had published several reports describing the fossils and age of the new species *Australopithecus anamensis* from Kanapoi and Allia Bay. But Pickford had doubts about the age and identity of the shinbone from Kanapoi, which was dated to older than 4 million years. If the shinbone was knocked out from under *A. anamensis*, then the species would not have a leg to stand on as an upright-walking hominid (although the teeth and other features made it a dental hominid). Geologists on Leakey's team, including Feibel at Rutgers, had confirmed that the sediments where the shinbone and other fossils were found were at least 4 million years old. But Pickford wanted to see the geology at Kanapoi for himself. He said later that he had an inkling it would stir things up. But he drove to Kanapoi anyway. On March 9, 2000, he set up camp under the trees near a dry riverbed. During a week poking around the arid gullies of Kanapoi, he found the conical piles of fine dirt that the Hominid Gang had sifted. He collected animal bones that he planned to subject to chemical tests to see if some were younger than others—to try to nail down whether the shinbone was younger than the other fossils of *A. anamensis*.

Word quickly reached the Leakeys that Pickford was exploring Kanapoi. On March 14, 2000, Richard Leakey wrote a letter to the director of the National Museums of Kenya, telling him that "at the moment, a Dr. Martin Pickford is in Turkana collecting fossils, some of which I believe he plans to export to France to study. Unless I am mistaken, the collection and exportation are illegal and actions should be taken without delay." Leakey suggested that the museum intercept Pickford, search him, confiscate his fossils, and press charges against him, according to a copy of Leakey's letter in Pickford's dossier on the case.

On March 17, Pickford packed up the camp, paid his workers, and drove right into a dragnet. As he stopped in the small market town of Kitale, he was met by some employees of the National Museums of Kenya that he knew. They told Pickford that the assistant curator of the National Museums' branch in Kitale wanted to meet with him. Pickford waited for ten minutes, sensing that it was some sort of trap. Sure enough, when the curator arrived, he was accompanied by an officer of the Criminal Investigation Division (CID)—Kenya's FBI. When Pickford told the CID officer he had a permit, the officer presented Pickford with the letter revoking his permit (the same letter dated to November 2, 1998, that Pickford thought was a forgery). Then the officer and museum employees searched Pickford's car, unpacking fossils he had collected from about ten areas at Kanapoi, mixing up their labels in the process of confiscating them and moving them to a museum truck. Pickford was held overnight at a hotel on the way to Nairobi while the CID officer slept in his car. The next night, Pickford was booked and moved to a communal cell at the Kileleshwa Police Station in the Nairobi area, where he spent several sleepless nights with a dozen other men. In the meantime, Gitonga called dozens of police stations searching for Pickford, including the Kileleshwa jail; Gitonga was told Pickford was not there.

On his third day of being held, Pickford was charged with the illegal excavation of fossils. After five days of incarceration, Pickford was released on March 21 after posting bail. He held a press conference where he called his arrest an intimidation tactic, and flew back to Paris as scheduled. The next month, Kenya's attorney general dropped the case, declaring it "nolle prosequi." Pickford and the Community Museums of Kenya then sued the Kenyan government, the National Museums of Kenya, and Richard Leakey for false arrest, seeking $1 million in damages. The case has been called to court from time to time for discussion—but it has been postponed each time (due to a heavy backlog of cases) without resolution.

To no one's surprise, the Tugen Hills were not big enough for both teams. Pickford and Senut were interested in the same fossil sites as Hill— those sediments that represented key periods in human evolution. Hill complained that Pickford dug trenches through a site where his team had done a carefully controlled excavation, thereby damaging their study of the ancient environment and possibly fossils. John Yellen, the director of the U.S. National Science Foundation's archaeology program, which funded Hill's team, would say it was not good for science to have two teams working at the same site, without cooperation: "The fossils lose an enormous amount of their value unless put in the proper stratigraphic and chronological context."

Pickford and Senut also found that they didn't like encroachers on the land where they had worked—or intended to collect fossils. In September 2003, Senut visited a small fossil site called Sagatia that Pickford had found in 1973, only to find that the Yale team already had been there earlier that summer, in June. As she came across piles of excavated dirt and shafts of bone, she fumed that the Yale team had damaged the fossilized bones and teeth of animals that she would have collected. She was particularly upset because she had hoped that her team's Kenyan graduate student would be able to study the animal fossils at Sagatia. Now that would be impossible.

Senut and Pickford complained to Gitonga. He, in turn, claimed that the Yale team had "proceeded to raid a fossil site" in a letter he wrote on September 29, 2003, to the dean of the graduate school at Yale University. He accused Hill's team of having collected hundreds of fossils of animals and reburied them, creating "a disaster of monumental proportions." He then threatened legal action against Yale University for damaging the fossils. He sent similar ominous letters to the National Science Foundation, again threatening to sue, and to paleoanthropologist Clark Howell at Berkeley, who was the lead scientist on an initiative that collected published data from seventeen different teams, including Hill's, to compare data from different sites from the late Miocene of Africa.

Hill's acid response was to point out that Sagatia had been part of his team's original permit area from the National Museums of Kenya for many years and was a site he had worked at years earlier. If anyone had the right to complain about their site being raided, it was he and other members of the Baringo Paleontological Research Project. One of his Yale graduate students, Katie Binetti, had been comparing the fossil animals from Sagatia with fossils from Tabarin, where the hominid jawbone had been discovered years earlier. She had conducted what is called a "total surface collection," collecting all the fossils on the ground surface that were identifiable (but not broken shafts) to get a sense of the ratios of different types of species at the sites. The idea was to see if there were biases toward one type of animal or fossil in the collection process and to comb the site thoroughly for hominid bones.

Gitonga would not be content with sending letters of complaint to American scientists and funding agencies. On the morning of December 3, 2004, he walked into Binetti's camp at Rondinin in the Tugen Hills to personally evict her. It was the same site where Pickford had camped ever since 1973, and Gitonga felt territorial about it, even though Hill had also camped there years earlier. Binetti had been granted a Fulbright-Hays fellowship to work in Kenya for the year, and she was there with a small crew of employees from the National Museums. One of these was Boniface Kimeu, her project manager, who is the son of the legendary Kenyan fossil hunter Kamoya Kimeu, who had worked with the Leakeys for many years. Another graduate student from the University of Michigan was visiting the camp that week. But none of them were in camp when Gitonga arrived at about ten A.M., so he told the camp cook instead that he was there to evict the team and to confiscate their fossils. The cook sent for Binetti, who was working at Sagatia that morning.

When Binetti returned to the campsite with Kimeu and the graduate student from Michigan, Gitonga was waiting for her. He had a gun tucked into the waistband of his trousers, but it was clearly visible, according to Binetti. Gitonga told her he was there to deal with her in the way he dealt with criminals, because she was trespassing and did not have valid permits to be at Sagatia. Binetti, who was twenty-eight years old and from Los Angeles, was appalled that she was being confronted by an armed man she had never met before over access to a research site. But, not to be deterred, she quickly regained her composure and got her research permits out of her tent to show Gitonga. Gitonga told her that one of her permits was not valid (it was), and told her that the director of the National Museums—Idle O. Farah—didn't know she was there (Farah did, and Binetti produced a letter from him, as well). Gitonga then claimed that he had an agreement with the clan that owned the land where she was camped.

As Gitonga persisted, Kimeu went to get the crew of twenty-seven people that Binetti had hired from three local villages, who were still working at Sagatia. When they arrived back in camp, many were angry that Gitonga had a gun. They also protested that he did not own the land at the campsite— and that another person with Gitonga did not own the land at Sagatia, as he claimed. Neither one could produce title deeds for the land. At that point, Binetti thought Gitonga would not use the gun in front of a crowd, and she started taking photos of him with the gun tucked into his pants. Gitonga backed down, but he said he would be back the next morning for a meeting with the local chief and his two assistants to settle who had the right to work in the area. After Gitonga left, Binetti drove several hours to the district offices in the town of Kabarnet to file a police report. She returned with two armed police officers, who stayed in her camp that night.

The chief and his assistants showed up for the meeting on Saturday morning, December 4, 2004, as promised. Gitonga did not show up. In his place he sent a local man who worked with the Community Museums of Kenya, Samuel Talam, to give the chief a copy of his permit to work in the "Mabaget formation," which is Pickford and Senut's name for the sediments at Sagatia. The chief and his assistants were angry that Gitonga did not show up, and after looking at Binetti's permits, they told her she had the right to work there. Although she had already received permission in November from local officials to be there, in addition to her permits, she was pleased to get the reassurances from the chief and his assistants. But she was unnerved later when the police questioned Talam and confiscated a paper from him that turned out to be a record of Binetti's and her team's movements during the entire time she had been in the camp at Rondinin.

A strong local reaction erupted as word spread of Gitonga's threats to a young graduate student. Not only were the local people angry that he had a gun, but the new director of the National Museums of Kenya, Idle Omar Farah, would renew his efforts to stop the Community Museums from issuing permits for research in the area. He also promised to travel with Binetti to the Tugen Hills to tell the chief and the local authorities in person that she had his permission to be there. Binetti returned to work in the Tugen Hills in 2005, but she found the episode deeply troubling. "As far as I know, no other graduate student has to deal with this kind of safety issue at a fossil site," she said.

She was also concerned about the safety of her team members. "Some teams have armed escorts because of conflicts between the local people. But this is somebody targeting me and my field crew for the research I am doing. I think the scientific community should know about this."

The conflict in the Tugen Hills and the specter of two teams prospecting the same deposits was destabilizing for paleoanthropologists working well beyond the Baringo basin. It was a violation of one of the tacit rules of paleoanthropology—and many nations' antiquities laws; one does not collect fossils on another team's claim if that team has a permit and is still working at the site. If this could happen in Kenya, where there were better regulations than in most African nations, it could happen anywhere. A wave of paranoia rippled through researchers' camps from Chad to Ethiopia to Tanzania.

The traditional ways of doing paleoanthropology were changing, and the unstated rules were being ignored. Gitonga called it "liberalizing" multiple sectors of the government and public life, including the museums and the management of research permits. Hill called it "undermining really good Kenyan laws on antiquities and monuments." Yellen at the NSF would observe that foreign researchers, including Hill and his American-funded paleoanthropological team, had no choice but to abide by the decisions of the Kenyan government regarding the Tugen Hills. But after six years, the Kenyan government still had not made any decisions, and both teams were still fending for themselves.

CHAPTER THIRTEEN

TOEING THE LINE

Do not steal another person's site, particularly when that person is
a local scholar in a developing country.

TIM WHITE, 2000

On April 13, 2000, an Ethiopian graduate student made an announcement that stunned paleoanthropologists at a meeting in San Antonio, Texas. It
was 8:15 A.M. when the student, Yohannes Haile-Selassie, took the podium to
give a report on his discovery of a new fossil site in the Afar depression of
Ethiopia. More than one hundred researchers had shown up for the early lecture, because Haile-Selassie (who is not related to the former emperor of
Ethiopia) was quickly becoming known as a researcher to watch. At thirty-
nine, he was slender and soft-spoken, with a low-key affect. Like many
Ethiopians, he was modest about his achievements. But there was no question
that Haile-Selassie was formidable in the field—he had found the first fossils
of the 4.4-million-year-old skeleton of *Ardipithecus ramidus,* and rumors were
circulating that he had discovered another early human ancestor that was a
million years older.

So when Haile-Selassie predicted that a new area called the Mulu basin
was going to become one of the best paleontological sites in the world, researchers in the audience paid attention. Then Haile-Selassie flashed slides of

three teeth of early human ancestors that he had found working at a site he called Galili. They were about 4 million years old, which made them some of the oldest teeth ever found—and probably the first Ethiopian fossils of *Australopithecus anamensis,* the species Meave Leakey and the Hominid Gang had found at Kanapoi. But what jolted the early-morning audience wide awake was Haile-Selassie's statement at the end of the talk that a team of foreign researchers had "claim-jumped" Galili two months earlier, in February. He did not name the members of the team, but several were in the audience, sitting in the dark.

When the lights came up, one of the researchers, Gerhard Weber of the University of Vienna, immediately stood up and asked to speak. Tall, with a shock of spiky black hair and an Austrian accent, Weber walked up to the podium, where he protested that his team had the proper permits to work in the Gedamaitu region, which included Galili. He was a junior member of the team led by the prominent Austrian anthropologist Horst Seidler, best known for his work examining the Iceman "Ötzi," the fifty-three-hundred-year-old frozen mummy discovered in the melting ice of the Alps in 1992. Seidler was not at the meetings, but a senior American anthropologist, Dean Falk, who had worked with the Austrians while they were working there, stepped in to back up Weber. She produced xeroxed copies of permits to the area later and said that this was all a misunderstanding. The room crackled with tension, as researchers tried to figure out what had happened. A few grumbled that a scientific lecture at the annual meeting of the American Association of Physical Anthropologists was not the place to air disputes about fossil sites.

After the talk, Haile-Selassie and other members of the Middle Awash research project (including White, who was not at the meetings) were indignant—if they could not raise this issue at the professional society for anthropologists, where else could it be broached? This was not the first time an Ethiopian graduate student had claimed at the annual anthropological meetings that foreigners were encroaching on his site—just a few years earlier, in 1995, an Ethiopian graduate student named Sileshi Semaw had accused American paleoanthropologist Don Johanson and his team of encroaching on a fossil site called Gona, where Semaw held a permit. It was right next door to Hadar, where Johanson had found Lucy, so the dispute boiled down to both teams holding overlapping excavation permits for the same area. But with more than one allegation of hostile takeovers of sites in Kenya and Ethiopia, these Ethiopian researchers, in particular, were on the defensive. They thought it was high time that senior scientists in the professional society made it clear that it was unethical to search for fossils at other researchers' sites. But to say that paleoanthropologists are a highly independent lot is an understatement—many students train through an apprentice system, learning their science and behavior in the lab and field from their individual mentors. As a result, they often forge alliances to their mentors and university, forming intellectual clans that give new meaning to the term "schools of thought." The notion that paleoanthropologists would voluntarily adopt a code of ethics or formally offer guidance on best practices in the field, in the manner of graduate programs in law or business, was as remote a possibility as getting paleoanthropologists to agree on how to arrange fossils on the human family tree.

The politics of doing research in developing nations also meant that every paleoanthropologist in the audience at the meetings knew that nothing was straightforward in any of these cases; in each situation, the paleontological arrivistes could produce permits to do research in the disputed terrain. And every single team vying for access to fossil sites had friends—and, in some cases, lawyers—in high places in the governments of the nations where they worked. The underlying question in each case was, Were the teams that moved in on other people's territory—and the bureaucrats who gave them the permits to the same sites—genuinely confused about the boundaries of the fossil fields? Or did the team leaders knowingly take advantage of government bureaucrats and local politicians or tribal leaders? In the absence of conflict-of-interest regulations in these nations, was it wrong to curry bureaucrats' favor with promises of coauthorship on publications, trips to hand-carry fossils to Europe or the United States, invitations to lavish embassy parties, and promises to build museums, schools, or clinics? Were those promises genuine attempts to build infrastructure and to train Africans, or were they superficial offers to gain access to prime turf? And did the teams seeking access to sites knowingly obfuscate the work done by the first researchers to work there?

In the case of Galili, the Middle Awash project members would suggest that Seidler knew he was moving onto Haile-Selassie's site. But in a letter to *Nature* shortly after Haile-Selassie's allegations in San Antonio, Seidler would protest vehemently that he had obtained his permit legally and transparently. He said it was an insult to him and to Ethiopian bureaucrats to suggest otherwise. He also insisted, "On my word of honor, I never knew it was Yohannes's site."

No one disputed that Haile-Selassie had discovered Galili in 1997. He had found fossils there with Ethiopian paleoanthropologist Berhane Asfaw, who had invited him to explore the central Afar rift zone with him. Both worked with White in the Middle Awash, but Asfaw had a permit to conduct a survey of the entire region for the Afar regional government. Using satellite maps and aerial surveys, Asfaw and Haile-Selassie zeroed in on the Mulu region, which is about 300 kilometers northeast of the Ethiopian capital of Addis Ababa. They scoured an area ten kilometers off the road, mainly on foot, and discovered a diverse range of animal fossils at Galili. At 3.6 million to 4 million years, the fossils at Galili were the right age to compare directly with the fossils of animals found at the sites farther north in the Middle Awash where their team had found *Ardipithecus ramidus*. Haile-Selassie planned to describe the fossils of mammals from Galili as part of his doctoral dissertation, and he also planned to return to Ethiopia to build his career on work at this study area.

Haile-Selassie returned to Galili in January 2000 to search for more fossils. In February 2000, he heard that foreigners were there, and he returned to Galili, with a letter saying it was his research area. Indeed, he found a large team of international researchers camping two hundred meters from the site where he had found the teeth, with a cohort of armed guards from the Issa tribe. In a confrontation on the spot with Seidler, he protested that they were working on his site. In fact, he realized that he had even left tire marks in the dirt the month before, leading right to the spot where he had found the teeth, which was also obvious from the piles of silt he had sifted. Seidler told him he had a legal permit to the area, and tried to ease the tension by inviting Haile-Selassie to work with his team.

Haile-Selassie would have none of it, protesting that the offer was "completely inappropriate, as the site was mine in the first place." He returned to Addis Ababa, where he met the director of the Authority for Research and Conservation of Cultural Heritage, who told him that Seidler did indeed have a permit for the Gedamaitu region, including Galili. Seidler had sought a new permit to the area in January 2000, after he had gone on a short trip weeks earlier to survey the region. He had found that his original permit covered a rocky area with no fossils a few kilometers south of Galili. The local people told him there were fossils in the Galili area. After his survey, he wrote to American geologist Jon Kalb, who had lost his permit to the Middle Awash when he was forced to leave Ethiopia in 1978 amid charges he was a CIA agent. Kalb looked at a map and told Seidler to apply for a new permit for another area, which included Galili. Shortly afterward, a cultural minister approved Seidler's permit for that region.

Haile-Selassie protested, reminding the cultural heritage official that he had discovered Galili, and telling him it was in the Mulu basin, where he held the permit. He had also filed reports annually with the same official, detailing his discoveries at Galili. But to his great frustration, on April 9, 2000, the official overruled his objections and renewed Seidler's permit for three years for the same area, including the Galili region. Haile-Selassie sent a letter to *Nature*, complaining that Seidler had kicked him out of his own site. He included a photo of himself at Galili in 1998, and a photo of Seidler's team camped right next to the same spot in 2000.

Seidler did not back off, however, countering that as White's student Haile-Selassie had access to large areas of the Middle Awash and suggesting that he didn't need Galili as well. In fact, members of Seidler's team saw Haile-Selassie not as a defenseless Ethiopian graduate student but as a mem-

ber of the powerful Berkeley team with White in the role of adviser, or-
chestrating strategy from Berkeley. It was clear to all that this was no ordinary
academic dispute but intense warfare between powerful factions headquar-
tered in Vienna and Berkeley—and that Seidler had the support of more
than one American who was angry with White. Ethiopian cultural heritage officials then gave Haile-Selassie another blow. They approved new
regulations controlling physical anthropology projects that included a time
limit of three years to hold research permits to an area. By that standard,
Haile-Selassie's time at Galili had run out, and he was unable to renew the
permit.

White and Ethiopian geologist Giday WoldeGabriel, who was one
of the leaders of the Middle Awash research project, stepped in to complain about the new regulations to high-level Ethiopian government officials—including the prime minister and foreign minister of Ethiopia when
they were on a state visit to Washington, D.C. In addition to time limits on
sites, the new regulations nationalized research equipment, such as vehicles,
even if they were bought by foreign governments. They also shrunk research
areas to ten kilometers and prevented geologists from working for more than
one project.

White's high-powered intervention did not sway the Ethiopian authorities. The field permits for White and the Middle Awash research project were
suspended in February 2001—a year after the Austrians claimed Galili—and
then revoked later without explanation. U.S. National Science Foundation officials got involved in March 2001, meeting with the Ethiopian ambassador in
Washington to discuss their concerns about the new regulations. But the Middle Awash research project missed an entire season of fieldwork in the fall of
2001. White and his colleagues, including Haile-Selassie, would regain their
permits to the Middle Awash in 2002. But Haile-Selassie would finally give up
trying to regain his permit to Galili, leaving it to Seidler's team. He brokered a
deal instead where he would rescind his claim to Galili if instead he could have
a permit that allowed him to explore a large area, almost the size of Connecticut, north of the Middle Awash research area. He didn't know it then, but one

site alone—called Mille-Weranso—would prove an even better locale than Galili. Unintentionally, Seidler ended up doing Haile-Selassie a big favor by diverting him from Galili.

⋙

In the midst of the battle over Galili, White wrote an incendiary article published under the tame title of "A View on the Science: Physical Anthropology at the Millennium" in the *American Journal of Physical Anthropology*, the journal of the professional association. It was one of a series of opinion pieces written by senior scholars giving friendly advice to young researchers. But, in truth, it was White's manifesto. He aimed it squarely at his peers, who would call it "The Tim Commandments." It included the following recommendations:

> In the field, do not think you're going to step out of the vehicle and pick up a hominid within 20 meters on the first day. Do not claim that you found the fossil if the other person did. The truth will eventually, some day, come out. Do not purchase fossils. Do not bribe officials. Do not steal another person's site, particularly when that person is a local scholar in a developing country. . . . In the armchair, do not assume that field workers "stumble" over fossils or are rendered stupid by getting dirty and absorbing lots of ultraviolet light. Sometimes they can even correctly interpret what they find. . . . Do not try to revise history to give yourself an advantage, or make yourself look smarter or better than other workers. . . . Modern paleoanthropology is predicated on teamwork, rather than individual success. Do not let your ambition distort your ethics.

It ended with the observation that "the science of paleoanthropology at the millennium is in serious trouble. The paleoanthropological commons are at risk from the selfish activities of practitioners. Tenure-tracked, mediaphilic paleoanthropologists seem unlikely to rescue the discipline from tragedy. They simply have too much to gain by acting individually and institutionally, rather than for the common good." It was an astute, if pessimistic, assessment of the state of the field in 2000.

~~~~

The same year that Haile-Selassie and Asfaw discovered Galili, in 1997, Haile-Selassie made another discovery in the Middle Awash that would eventually prove to be much more significant than the teeth at Galili. Back in 1995, White and Asfaw had decided that Haile-Selassie was ready for a new challenge. Impressed with his talent for finding fossils in the Middle Awash, they had decided that it was time to give him some independence—and to send him into the most hostile and remote terrain in the southern Afar rift. They would ask him to lead the effort to survey the dry washes and steep slopes of the western margin, a series of ridges that rises west of the Middle Awash. White and Asfaw couldn't promise Haile-Selassie that he would find hominids, but they knew that he would find something interesting for his Ph.D. dissertation. Jon Kalb had surveyed parts of the region years earlier, and White and Asfaw had surveyed the area briefly in 1981 and, several more times, in the early 1990s.

Haile-Selassie did not let them down. He had been waiting for years for this kind of break. Originally from the province of Tigray, in northern Ethiopia, he had moved to Addis Ababa when he was nine years old to live with an uncle who'd spotted his potential as a student, and sent him to school there. Unfortunately, his uncle died when he was in ninth grade, and did not see him admitted to Addis Ababa University, where he earned a degree in history in 1982. He was one of only four students out of forty-two not assigned by the government to be a teacher—instead, he was offered a position at the Ministry of Sports, Youth, and Culture, along with Sileshi Semaw, now a paleoanthropologist at Indiana University, and Yonas Beyene, now an archaeologist at the Authority for Research and Conservation of Cultural Heritage in Ethiopia. Semaw and Beyene, who still work closely with Haile-Selassie and other members of the Middle Awash project, were assigned to work at the National Museum, where foreign researchers soon offered them scholarships to earn Ph.D.s in the United States and France.

Haile-Selassie also wanted to win a scholarship to earn his Ph.D., so he transferred to the National Museum. He would have to wait for his break: the eight-year moratorium on doing anthropological or archaeological fieldwork in Ethiopia had put a damper on scholarships. Finally, in 1989, when Asfaw returned to do his Paleoanthropological Inventory of Ethiopia, he invited Haile-Selassie to join him. After a few years of working in the field, Haile-Selassie told Asfaw and White that he wanted to earn a Ph.D. White advised him to submit an application for graduate studies, and by the fall of 1992, he was in Berkeley taking classes. In 1995, he was looking for a topic for his Ph.D. dissertation in anthropology when White and Asfaw asked him if he wanted to work in the western margin.

Off and on over the course of four years, Haile-Selassie and senior geologist Giday WoldeGabriel, who volunteered to help with the geology, would leave the main team working around Aramis and for a week or so at a time would explore the western margin. They were well suited to surveying the remote terrain together. "We knew that the area was remote and difficult to work, but we were sure that if there were hominids there, Yohannes and Giday would be the ones who could find them," White said. The two would drive as far as possible into the western margin, then put up a tent beside their car so they could get an early start the next day, heading out on foot with backpacks. Sometimes they walked as much as twenty or thirty miles, following goat and camel trails through ancient lakebeds and over prehistoric lava flows or bushwhacking through thorny brush. The pair was seeking the bare patches they had targeted from aerial photographs and space images—the basins and slopes with ancient sediments exposed by tectonic forces or by the torrential rains that come along twice a year. They were so far off the beaten path that, more than once, they surprised Afar villagers and found themselves looking down the wrong end of a gun barrel. Haile-Selassie quickly learned enough of the Afar language to explain why he was exploring their territory, and he talked to the local chiefs to get their permission to work there. Eventually, the local Afar would speak of him with respect, and ask about him whenever other members of the team were there without him.

On the morning of December 16, 1997, after Haile-Selassie had been working in the western margin for two years, he spotted a piece of jawbone lying among the basalt cobbles on the ground. Intent on collecting fossils of animals for his dissertation on the ancient environment, he was not even thinking about finding hominids. He had been working in the Alayla basin for just five minutes that morning when he picked up the bone and saw a familiar-looking molar set in it. He called White on his mobile radio and told him he had found a hominid. White asked, "How do you know it's a hominid?" Haile-Selassie responded, deadpan: "Because it's a hominid." Wolde-Gabriel, who was showing White some of the geology of the area, recalled walking very quickly to reach the spot a mile away where Haile-Selassie had found the fossil. When they got there, White agreed—it was the jawbone of a hominid.

White would comment later on Haile-Selassie's uncanny ability to find fossils of hominids, describing it as a talent that has to do with the recognition of three-dimensional shape—a talent that is innate. In an interview with a reporter for the Cleveland *Plain Dealer* in 2004, he said:

> Just like playing a piano or pitching a baseball or anything else in life, some people really have that talent and some people don't. If I gave you a tooth and said what is it, you could tell me it was a tooth but you wouldn't be able to immediately tell me it's a rhino tooth. A week later you find another tooth and if you didn't have these skills, you'd say well maybe it's a bear, maybe it's a sloth—I don't know what the hell it is. But Yohannes would look at it and say, "That's a rhino tooth." Right away. It's this anatomical recognition skill that is absolutely essential in the field, because the fossils that you find aren't nice intact skulls and jaws. They're usually broken in very small pieces that are very hard to identify, and you find literally hundreds of thousands of such pieces every day. You've got to be able to tell which one's important

and which one's not. What that means is you've got to discriminate the human ancestor among the hundreds of other animal species whose bones and teeth you're finding scattered across the same outcrop.

In fact, Haile-Selassie did not realize how special the jaw fragment was until he was in the lab, analyzing the fossils, almost a year later. As he compared the jaw fragment and four other teeth the team had found, as well as arm and hand bones, with fossils of Lucy's species and *Ardipithecus ramidus*, he realized that he had found something even more ancient and apelike. When geochronologist Paul Renne told him that the fossils came from sediments that were 5.8 million years old, based on argon-argon dating, Haile-Selassie knew he had found the remains of the most ancient member of the human family then known—one that was 1.4 million years older than *Ardipithecus ramidus*, from Aramis.

He did not have the right fossils to determine if they were an early form of *Ardipithecus ramidus* or whether they belonged to some new type of human ancestor. He remembered the controversy that had swirled around Patterson's partial jawbone found at Lothagam in Kenya three decades earlier and the lessons of initially diagnosing *Ramapithecus* as a hominid on the basis of a partial jaw fragment. He decided to wait to publish his discovery until after he had more data—in the form of more teeth and bones that would help prove that the fossils were the remains of an ancestor of humans, not chimpanzees.

Over the course of the next four years, Haile-Selassie and his colleagues fanned out over the western margin, collecting a total of eleven hominid fossils from five individuals and twenty-five hundred fossils of animals, which underscored that the hominid fossils were just tiny nuggets in an immense landscape littered with pebbles, boulders, and animal bones. Those eleven fossils ranged in age from 5.2 million to 5.8 million years. They offered a partial portrait of an ancient primate, from its teeth down to its toe—literally. A young biological anthropologist named Leslea Hlusko had found a 5.2-million-year-old toe bone at Amba, near the site in the Middle Awash where the original *Ardipithecus* had been found.

The teeth convinced Haile-Selassie that he had a very early hominid, because they looked like they were in transition, between teeth similar to a chimpanzee's and those of an ancient hominid. The lower jaw was roughly the same size as a common chimpanzee's, but the back teeth were larger and the front teeth narrower, suggesting that the creature ate less fruit and soft leaves and more fibrous food than a chimpanzee. He did not have the all-important upper canine whose size and sharpened—or unsharpened—state could tell him if this was a dental hominid. But he had lower canines, and they were diamond-shaped in cross section, showing that they wore down from the tip, as in later hominids, in contrast to the inverted V shape of lower canines found in chimpanzees.

Finally, the younger toe bone from Amba suggested that its owner had walked upright, because the joint was slanted in the same direction as the toe bones of later hominids, including Lucy. What Haile-Selassie and locomotion expert Owen Lovejoy saw in the toe bone was enough to convince them that this new hominid could toe off, pushing forward by leaving the front part of the foot on the ground and lifting the heel—anatomy seen in hominids but not in chimpanzees or other apes, which walk by pushing off the outside of the feet. The new hominid had a toehold, so to speak, on the lineage leading to modern humans.

At the beginning of 2000, only seven fossils of extinct apes and hominids were known from the critical time in the late Miocene when the earliest hominids arose. Haile-Selassie went to the field in late 2000 after he submitted a paper to *Nature* in October, announcing the new fossils from the western margin that would be the earliest known human ancestors. In the paper, which was being reviewed by his peers, he had identified the fossils as a new subspecies of the root ape, which he called *Ardipithecus ramidus kadabba*. Haile-Selassie used the word *kadabba,* which in the Afar language means "basal family ancestor," to honor the local Afar people. Translated from Afar *(Ardi* and *kadabba)* and Latin *(pithecus* and *ramidus),* this would be the basal-family-ancestor root ground ape. Later, after finding more fossils, he would rename it *Ardipithecus kadabba,* to indicate it was a new species. But when

Haile-Selassie returned from fieldwork in the Middle Awash with White in early 2001, they learned that there had been a press conference in December 2000 in Nairobi announcing the discovery of an even older hominid. "It was bad," recalled Haile-Selassie several years later. Not only had he lost Galili in 2000, but he also had been scooped on his announcement of the earliest known hominid.

CHAPTER FOURTEEN

MILLENNIUM MAN

I have weathered the storm. I have beaten out my exile.

EZRA POUND

In this world there are two tragedies. One is not getting what one wants, and the other is getting it. The last is much the worst.

OSCAR WILDE

On the morning of November 4, 2000, Martin Pickford and Brigitte Senut returned to the Tugen Hills of Kenya for the first time since Pickford's arrest. They jostled in their battered Toyota truck down a steep switchback road that their team had rebuilt for cars—a dirt road they called "the Limit" that would link the museum they were erecting on top of the ridge with the dirt tracks in the valley below. It was a far cry from the days in the 1970s when Pickford would hike into the Tugen Hills with a backpack, and the local people would call from ridge to ridge to tell his friend and assistant, Kiptalam Cheboi, that he was coming. That custom died out long ago, but on this morning, Kiptalam was expecting them anyway when they pulled into their camp, set beneath the shade of tamarind trees alongside a dry river. Life is quiet in the Tugen Hills, except when the contentious Pokot tribesmen from the north slip in under cover of darkness and raid the Tugen cattle at night. Word travels quickly when strangers arrive in the area, particularly foreigners who are hiring workers to build roads and search for fossils.

As soon as Pickford and Senut arrived in camp at Rondinin, Kiptalam disappeared into his tent, then came out with a paper bag. When Pickford and Senut opened the bag, they were surprised to find two pieces of a jawbone with molars that looked small and square, like a human's. It was immediately obvious to them that the teeth belonged to a hominid, rather than an ape. There also was a finger bone that another team member had found nearby.

Just two weeks earlier, on October 21, Kiptalam had walked to camp by a different route than usual. (Kiptalam has worked with Hill and Pickford ever since he was a young man in the 1960s. Indeed, he is the same age as Pickford—fifty-six in 2000. A tall and slender man, he looked older after decades of working as a goatherd and fossil hunter in the Tugen Hills. Yet he still knew the crumpled landscape of the Tugen Hills better than anyone.) He was walking along an ancient streambed when he saw something sticking out of the red dirt on the ground at his feet. He looked closer and saw teeth. Two fragments of a jawbone were lying on the surface. Kiptalam said his whole body was filled with excitement when he realized the teeth looked human; he could have jumped for joy. He collected the teeth and marked the spot with acacia thorns and a small cairn of rocks. (He'd learned long ago not to use small flags, because the local Tugen children collect them.)

On the day that Pickford and Senut arrived, he took them straight to the spot where he had found the fossils. It was a dry gully about eighty kilometers north of the equator in a remote area known as Kapsomin, where the hillsides are covered with prickly acacia shrubs, tamarind trees, and a species of heliotrope that emits a pungent smell. A few Tugen people live in thatched *bandas* on the ridges, where they have subsistence farms.

No sooner had Pickford and Senut begun to inspect the gully where Kiptalam found the two fragments of jawbone than Pickford got his second surprise of the day: he saw a bony ball sticking out of the red dirt. It was the head of a left thighbone. He gently eased it out of the earth and then, as he probed the same spot, he found part of the shaft of the thighbone still embedded in the ground nearby. He wrote in his field journal that night that the shaft of a "bipedal hominid" had a fresh break; he would say later that the thigh-bone was found in four pieces slightly separated from one another. Pickford recorded their precise location with a handheld global positioning satellite (GPS) device. He also drew a sketch of the site in his journal, noting that the fossils came from a layer of sediment that was at the base of the Lukeino for-mation, below a layer of basalt dated to 5.65 million years by Hill's team. "Therefore about 6 million years," he noted.

Pickford and Senut immediately recognized the significance of the dis-coveries. If the fossils really were 6 million years old, they would belong to individuals who were the oldest known members of the human family. They would predate by 1.5 million years the fossils of *Ardipithecus* from Ethiopia, published by the Middle Awash team six years earlier, and, possibly, the un-published fossils they knew that Haile-Selassie had discovered (although Pickford said he had no inkling they existed).

The excitement built over the next two weeks. The next day Senut found an upper arm bone nearby, and later team members discovered a canine tooth and more fragments of thighbones. By the end of two weeks, the team had found thirteen fossils from at least five individuals in the area, capping a remarkable field season in the Tugen Hills. When asked how they felt about the discovery, Pickford and Senut insisted that finding a hominid was no dif-ferent for them than finding a fossil of any other primate—or even a rat. Senut would say, "I find the complete skeleton of a small rat just as exciting as a hominid." And Pickford would say later that he felt sorry for people who studied only human evolution. "It's like reading a book and skipping to the last chapter to read the ending," he said.

Senut's initial reaction, however, revealed that finding the thighbone of the earliest hominid was not exactly like finding a rat. She had a sense of foreboding: "I told Martin to throw it in the lake. It would only bring us trou-ble." Pickford, too, would admit that he had nightmares for two weeks after he found the thighbone, worrying about whether the fossils really were 6 mil-lion years old. He asked himself whether it was possible that the fossils could have been younger but had been washed into the older sediments by water, since they were found in an ancient streambed. He was well aware that if he made a false claim, he would never live it down—"People would ruin us," he would say. But Pickford eventually ruled out the idea that the hominid fossils had been displaced, because the team was accumulating fossils of elephants, horses, hippos, and pigs that were all from extinct species that had lived in the Miocene, more than 5 million years ago. The bones were also all encrusted with a coating of limestone deposited by freshwater algae, which meant they had fossilized at about the same time and in the same place. And several had teeth marks from a big cat, such as an extinct leopard. This suggested that these hominid bones were the remains of a carnivore's meal that had been left in one place, perhaps beneath a tree, before they were covered in mud or water and fossilized. Pickford also had the advantage of having done his thesis on the geology of the area.

By the time the field season was over, in late November 2000, Pickford and Senut had little doubt about the age of the fossils they were finding—or their identity as hominids. They discussed the fossils with the Kenyan president at the time, Daniel arap Moi, who encouraged them to announce their discovery in Kenya rather than wait to publish it in a scientific journal, because it would "put Kenya on the map."

On December 4, 2000, they happily obliged Moi and held a press conference in Nairobi, where they introduced the fossils as the remains of a human ancestor they called the "Millennium Ancestor," since they'd been discovered in 2000. The fossils, they said, belonged to the earliest known member of the human family at 6 million years old, predating all other known hominids by 1.5 million years.

They held another press conference in Paris just three months later. With remarkable speed, they had submitted a paper describing the fossils and the geology of the site where they were found on January 16, 2001, to the journal of the French Academy of Sciences, called *Comptes Rendus de l'Académie des Sciences*. It was a surprising choice for the publication of a paper announcing the earliest member of the human family. But because the French journal required only two anonymous scientists to peer-review a paper, it would publish the discovery much faster than *Nature* or *Science*. Yves Coppens, who was a member of the French Academy, referred the paper to the journal for publication and was a coauthor. As a result, Pickford and Senut beat Yohannes Haile-Selassie to publication—Haile-Selassie's report announcing his fossils of *Ardipithecus ramidus kadabba* as the earliest known hominid at 5.2 million to 5.8 million years was still under review at *Nature* and would not appear for months.

When the paper on the Millennium Ancestor was published in March 2001, Pickford and Senut introduced it by its new scientific name: *Orrorin tugenensis*, drawing on the Tugen name and legend for the "original man" who settled the Tugen Hills. The name also had the aura of the French word *aurore*, or "dawn"—hence, dawn man. (Although Pickford would note that *Orrorin* also sounded like the Italian word for "horror," *orrore*, perhaps anticipating some rivals' responses to their discovery of the new fossils.)

Orrorin was more than an evocative name. It signaled that Pickford and Senut thought it was an entirely new type of human ancestor that belonged in its own genus. They based that claim on the anatomy of the upper part of the thighbone and the jaw fragments. In particular, they said that the head of the thighbone was large and "human-like" in relation to the size of the neck of the bone. They also pointed to a groove on the neck of the thighbone that was evidence of contact with a thick tendon in hominids that habitually walked upright. The bandlike tendon puts pressure against the bone when the hip is fully extended for a straight-legged stance or gait in upright-walking humans, leaving a furrow not found in apes. *Orrorin*'s small, thick-enameled molars also suggested it was a human ancestor, they wrote, because these were traits seen in later hominids. If so, it was an extremely early hominid, because Senut thought its arm and finger bones showed that it still spent time pulling itself up in the trees.

The naming of *Orrorin* as the earliest known hominid was not the only bold claim in the paper. Pickford and Senut also tried to rewrite early human history by kicking *Orrorin*'s closest competitors off the direct line leading to humans, starting with Lucy. Even though *Orrorin* was almost twice as old as Lucy, they wrote in their paper that its thighbone was more like a human's than those of Lucy and other australopithecines that lived much later—2 million to 4 million years ago. This meant that *Orrorin* walked in a more modern manner than Lucy did, Senut said. She had drawn a simple phylogenetic diagram showing *Orrorin* on the direct line to humans. A second line ran parallel to it, leading to Lucy and the other australopithecines and eventually to extinction. A third line repositioned *Ardipithecus* from Ethiopia right out of the human family. It showed *Ardipithecus* as the ancestor of African apes instead of humans.

It was a hypothesis notable for its simplicity and boldness—and its opposition to the prevailing view of human origins held by most paleoanthropologists. "This simple phylogeny contrasts starkly with mainstream ideas about human evolution, and glosses over many areas of controversy and uncertainty," wrote paleoanthropologists Leslie Aiello and Mark Collard, then of University College in London, in the journal *Nature*. It reminded some anthropologists of the view espoused in the 1960s by Louis Leakey, who had also thought that the australopithecines were not directly ancestral to modern humans. Senut, who had earned her Ph.D. studying the arm bones and locomotion of Lucy, had been the main proponent of that view ever since. But unlike Leakey, who would rely on differences in skulls and teeth to sort fossils, Senut reasoned that the skeleton was more reliable for figuring out evolutionary relationships. If bipedalism was the defining trait of being an ancestor of humans rather than African apes, she would look to the bones that had been rearranged to walk upright to identify the direct ancestors of humans.

This perspective, however, meant that Senut arranged the fossils in the human family tree differently than most other researchers did. Long before she and Pickford discovered *Orrorin*, Senut had proposed that there were two types of early hominids, defined by their anatomy from the neck down. Those that still spent a great deal of time hanging out in trees and that walked upright with a bent knee were australopithecines, and, in her view, they—including Lucy—went extinct. But those that walked upright in a more modern manner she placed in a preexisting genus called *Praeanthropus,* which she considered ancestral to modern humans. She included a few fossils that had been attributed to *A. anamensis* and to Lucy's species—but not Lucy—in that genus. With the discovery of *Orrorin,* she now had a fossil with a modern-looking thighbone that fit her prediction for an ancestor of *Praeanthropus* and modern humans.

But few paleoanthropologists agreed that there were two groups of such early hominids defined by differences in their lower limbs. In their report in *Nature* entitled "Our Newest Oldest Ancestor?" Aiello and Collard warned, "These are exciting times in the study of human origins. But excitement needs to be tempered with caution in assessing the claim of a six-million-year-old direct ancestor of modern humans." They concluded (without seeing the fossils): "The age of *Orrorin* undoubtedly makes it a highly important addition to the debate about human origins. But we are a long way from a consensus on its role in human evolution."

Pickford and Senut, however, appeared to revel in the role of outsiders challenging the establishment. They were beaming at their initial press conference. Pickford exulted that *Orrorin* showed paleoanthropologists the opening chapters of human history. Their fortunes had changed dramatically in the nine months since Pickford was arrested. Pickford wrote in his field journal on March 21, 2001, "A year ago today, I was in front of the judge, charged with collecting without a permit! All in all a good year for me and a bad one for REL [Richard Leakey]."

Andrew Hill was at home, asleep, in the predawn hours of December 4, 2000, when he got a telephone call from Nairobi. It was the morning that Pickford and Senut held their press conference in Nairobi, and the caller claimed to be a reporter from the BBC. He wanted Hill's reaction to the discoveries in the Tugen Hills. This was when Hill first learned that Pickford and Senut had made an important discovery in the Tugen Hills where he had worked—and where his team still held the research permits from the government of Kenya. He found it particularly galling to hear that Kiptalam had found the jawbone, since he had worked for Hill for many years as part of the Baringo Paleontological Research Project.

His response to the call was to ask, "What did the National Museums of Kenya say about these discoveries? Were they legal?" The caller had such an odd reaction to his questions, however, that Hill wondered if he was a legitimate reporter—or someone calling to annoy him. He never did find his comments in BBC accounts of the discovery.

Hill was not the only paleoanthropologist unsettled by the new announcement. Paleontologist Alan Walker, who was the codiscoverer of *A. anamensis* and who had analyzed a thighbone of another, more recent hominid, would tell the *New York Times* he was "not impressed with their evidence" for bipedalism. Paleoanthropologist Carol Ward of the University of Missouri in Columbia, who was an expert on lower limbs and who had worked on *A. anamensis,* complained that Pickford had the wrong end of the thighbone—the knee would provide better evidence of upright walking. Both had worked closely with the Leakeys, so Pickford would chalk up their negative responses to politics.

But others who had no link to the Leakeys would also express reservations about the claims made for *Orrorin* as a hominid. Aiello and Collard wrote in *Nature* that the features that suggested that the teeth belonged to a human ancestor—small molars and thick enamel—could be adaptations to a similar diet, rather than evidence of shared ancestry. Others would criticize embarrassing errors in the anatomical traits described in the manuscript—such as calling the groove on the thighbone the intertrochanteric groove, which it was not (it was the obturator externus groove), and reporting an incorrect measurement for enamel thickness (Pickford said it was a misprint). Tim White would note later, "Given these irregularities, it is therefore not surprising that many readers were unconvinced of the accuracy of the first functional and phylogenetic claims made by Senut and Pickford about *Orrorin.*"

Although many questions were unanswered about *Orrorin,* White was among the first to take *Orrorin* seriously. He sent an e-mail congratulating

Pickford and Senut on the discovery of *Orrorin* as soon as he emerged from the Afar and heard about it. He visited Gitonga at the Community Museums of Kenya in Nairobi, but he did not ask to see the original fossils of *Orrorin*, which by then were stored in a bank vault in Nairobi (although Pickford and Senut would show casts of *Orrorin*'s teeth to White's colleague Gen Suwa, in Paris). Later, White invited Pickford and Senut to join an international project to compare the ancient environments where the earliest hominids and Miocene apes were found. (Pickford and Senut declined because White did not consult the Community Museums of Kenya, Pickford said. Hill signed on to contribute his team's studies of the ancient environment of the Tugen Hills, but on the condition that Pickford did not.)

White also sent a series of e-mail messages to Pickford, asking questions about the fossils and, later, spelling out in detail how he thought the pair should analyze the thighbone, because he was completing an intensive study of a younger hominid thighbone. White recognized immediately that *Orrorin* was of enormous significance for understanding human origins, because it was the first well-preserved upper thighbone of a fossil hominid or ape older than *A. afarensis*.

If Meave Leakey, Alan Walker, and their colleagues were right that upright walking was already in place by 4.1 million years ago in *A. anamensis* at Kanapoi, then researchers had to look further back in time to see how the anatomy to walk upright was assembled in the first place—and to see if it looked different from the gait of australopithecines and modern humans. White and Haile-Selassie and colleagues were deciphering clues in the fossils of both types of *Ardipithecus* from Ethiopia. But as recently as the summer of 2005 they had not found an upper thighbone for either species of *Ardipithecus*. This made *Orrorin*'s upper thighbone at 6 million years particularly interesting. White wanted to know if *Orrorin* walked upright, and therefore qualified for true hominid status. He also wanted to know *how* it walked upright—with a straight, fully extended hip and leg like modern humans and australopithecines? Or did it use a new form of locomotion, such as pushing off the outside of its foot and swaying from side to side or some other unknown mode of walking?

The pelvis, knee, and ankle joints are best for diagnosing upright walking in a skeleton, but White and locomotion expert Owen Lovejoy had figured out how to tease an amazing amount of information out of an upper thighbone. At the time *Orrorin* was announced at a press conference, they were finishing a study of the thighbone of a 3.4-million-year-old teenage australopithecine from the Maka sands of the Middle Awash in Ethiopia. They had compared it with thighbones of more than 150 African apes and 250 humans to sort out which features in the exterior and interior of the thighbone were reliable for diagnosing upright walking. They had also pored over photographs, X-rays, and computerized tomography (CT scans) of the cross sections of thighbones of several fossil hominids, as well as those of thirty-five modern humans, chimpanzees, and gorillas. They concluded that several anatomical details, including the obturator externus groove that was seen in *Orrorin,* were reliable indicators of some form of habitual upright walking. They also confirmed that the pattern of bone distribution inside the neck of a thighbone was remarkably reliable at revealing how an animal balanced its weight and absorbed strain in its upper legs when walking.

The internal architecture of the thighbone shows the history of how the bone was used, because connective tissue is remarkably sensitive to strain. More bone is retained where there is more stress; in other words, the outer layer of the bone is more dense where it needs to be stronger. This is why humans are advised to work out with weights as they age, to prevent the loss of bone density. In humans, the neck of the thighbone is positioned at an angle that is 130 degrees from the vertical shaft of the thighbone. When humans walk upright, the strain and weight from above is distributed across the neck of the thighbone, which flexes like a cantilevered beam with support from only one side. But the bone is compressed by muscles and tendons that hold it in place, so more strain is felt on the underside of the bone, as if it were an arched stone bridge. As a result, the bone is thicker on the bottom, where there is more stress, and thinner on top, where there is less—a pattern not seen in apes that walk on all fours, which have an even distribution of bone density across the cross sections of their thighbones. White and his colleagues have also found that subtle differences in the distribution of the bone thickness can reveal differences in the way that a thighbone was used to walk upright.

With this in mind, in March 2001 White sent an e-mail asking Pickford if he had images of the interior of *Orrorin*'s thighbone. Pickford responded in April 2001, saying that yes, they had taken photos and measurements of the interior of the neck of the thighbone where it was broken before they glued two fragments together. Furthermore, they had brought the thighbone of *Orrorin* to a clinic in Toulouse, France, where they took CT scans of the 6-million-year-old thighbone, putting it under the scanner as if it were the broken leg of a patient.

White sent another e-mail to Pickford, suggesting that direct photos of the bone were always better than CT scans for measuring interior bone thickness. He included a prepublication copy of his paper on the Maka thighbone to illustrate his point. Pickford responded the next day, on August 6, 2001, saying that he would unglue the thighbone where it was broken across its neck and take direct photos of it when he returned to Nairobi. But at the end of the year, when White pinged him again, Pickford had changed his mind. He responded that the CT scans were sufficient to see the bone distribution—and that he didn't want to unglue the thighbone, because he didn't want to damage it.

Pickford and Senut and their colleagues published their CT scans the following year, in September 2002, in a scientific paper describing the thighbone in more detail. This time, they verified that the obturator externus groove was present, which made a strong case for *Orrorin*'s status as a hominid that used some form of upright walking. They also stated that the CT scans showed that the pattern of bone density was close to that of humans and australopithecines.

To White's immense frustration, no photographs of the cross section of the thighbone neck were included. And the CT scans were of poor quality and taken at the wrong angle to give a reliable measurement of the bone density, according to anthropologist James Ohman at Liverpool John Moores University in England, who was an expert on CT scans and coauthor of the Maka thighbone study. Ohman concluded that the quality was too poor to be reliable. White would send two more e-mail messages in 2002, urging Pickford again to take X-rays or unglue the thighbone to take photos—to no avail. White asked repeatedly why Pickford and Senut would not do this simple test—or reveal the results if they had done so.

Senut and Pickford protested that it was unnecessary to unglue the bone to photograph it and measure it, particularly if it could damage the fossil. They said the external features made it clear that *Orrorin* walked upright, so there was no point in breaking open the fragile fossil. Besides, Pickford would later add, the bone was broken in a zigzag pattern that made it difficult to photograph. Senut and Pickford also made it clear that they resented the pressure to produce unpublished data on *Orrorin* when they had yet to see images of the skeleton of *Ardipithecus ramidus.* In a fiery exchange at a meeting in Paris in September 2004, Senut turned the tables on White and challenged him to provide his own unpublished evidence about the mysterious skeleton he had been working on for almost a decade. But White pointed out that analyzing a brittle skeleton was far more arduous than studying one thighbone. Furthermore, he had never made the exceptional claim that *Ardipithecus ramidus* walked upright better than other hominids. "Exceptional claims demand exceptional evidence," he said.

Pickford and Senut did republish the CT scans, with corrections for the wrong angle. Their report in *Science* in 2004 with a pair of American researchers concluded that the scans showed that *Orrorin* walked upright. But there was no information on gait, and even though the CT scans had been adjusted with skill, they were still the same old scans with poor resolution, and, therefore, the team's conclusion that *Orrorin* walked upright better than Lucy was unsupported by the evidence. In a letter to *Science* in February 2005, White and Lovejoy said that the presence of the obturator externus groove did indeed suggest some form of upright walking and, hence, that *Orrorin* was a hominid. But "the adjustment of previously published data did not suffice" to prove that *Orrorin* walked like a human—or to answer the question of what type of hominid *Orrorin* represented.

Indeed, in the absence of definitive evidence in the form of X-rays, photographs, or new CT scans, many paleoanthropologists continued to harbor doubts about *Orrorin*'s status as a new type of early hominid. An increasing number accepted its status as a hominid. But questions persisted about the trustworthiness of the pair's methodology and analysis of the fossils, which came back to plague Pickford and Senut when they applied for research funding, and at scientific meetings. *Orrorin* remained in scientific purgatory, recognized as neither hominid nor ape. Andrew Hill, for one, did not include it in his new exhibit on human evolution at Yale's famous Peabody Museum. As Hill prepared to open the exhibit in 2004, he walked through it pointing out photos of the famous researchers who had found significant fossils of human ancestors in the past century, starting with Dubois and Dart and continuing on through the Leakeys to Tim White and Michel Brunet. There was no mention of Martin Pickford, Brigitte Senut, or *Orrorin*. He later explained, "I don't include *Orrorin* and Pickford and Senut, because I don't consider *Orrorin* yet to be an established taxon."

∼∼∼

In July 2001, just four months after Pickford and Senut published their report on *Orrorin*, Yohannes Haile-Selassie reported his own discoveries of the eleven fossils from Ethiopia that were almost as old as *Orrorin*. The toe bone of *Ardipithecus ramidus kadabba* (as well as the jaw, teeth, arm, and hand bones) came hot on the heels of *Orrorin*'s thighbone, prompting *Nature* editor Henry Gee to write in an accompanying article that "discoveries of fossil hominids are like buses: nothing for a while, then three come along at once." Besides *Orrorin* and *Ardipithecus ramidus kadabba*, a third, younger type of hominid had been announced earlier that year by Meave Leakey.

Even though the new fossils were a bit younger than *Orrorin*, many reports referred to *Ardipithecus ramidus kadabba* as the earliest known hominid, reflecting the uncertain status of *Orrorin*. The editors at *Time* magazine featured the discovery on the cover under the title "How Apes Became Human." Inside, the remains were described as belonging to a chimplike forest creature that "appears to be the most ancient human ancestor ever discovered." Much was made of the inch-long toe bone, which made it clear that *"kadabba* almost certainly walked upright much of the time."

But the evidence that the new hominid walked upright rested on a toe bone that had been found about twenty-five kilometers from the other fossils and that was about half a million years younger than the teeth and jaw fragment. In many ways, the evidence that *A. ramidus kadabba* had walked upright was no more water-tight than the evidence from *Orrorin*'s thighbone that it was bipedal. The new fossil evidence for both creatures was so scanty that it was impossible even to rule out that they were the same type of hominid, because there were few overlapping teeth or bones to compare directly. Haile-Selassie and his colleagues would suggest, in fact, that *Orrorin* might even be an early member of *Ardipithecus*, because the teeth of *Orrorin* looked a lot like the teeth of *A. ramidus kadabba*. (He had seen only photos, not original teeth, of *Orrorin*, but Suwa had seen casts.) They also were found in similar environments—Haile-Selassie's intensive study of the nineteen hundred fossils of animals from the western margin indicated that *A. ramidus kadabba* lived in a lush woodland. Pickford had also reported that *Orrorin* lived in the woods, suggesting that both of these early creatures were very primitive hominids still spending most of their time in the trees. If Haile-Selassie was right that they might be the same type of ground primate from the forest, then *Orrorin* would be reclassified as a new species or subspecies of *Ardipithecus*, since that name came first and, thus, took precedence over *Orrorin*.

To nobody's surprise, Pickford and Senut rejected that idea and kept *Orrorin* in its own genus. They cited a few differences in the teeth and other bones, such as difference in the thickness of enamel on the molars. But without more fossils to compare directly, such as upper thighbones or toe bones from both *Orrorin* and *Ardipithecus ramidus kadabba*, it would be difficult to discount the possibility that they were different versions of the same genus. And no matter what the fossils showed, the two teams might still see them differently. Where Pickford and Senut saw different types of hominids older than 4 million years—and evidence of a radiation or diversification of early human ancestors—White and Haile-Selassie saw two different types, at most. One editor caught in the middle—Henry Gee—would throw up his hands in print, writing in *Nature* in July 2001, "Sadly, I doubt that the status of these creatures can be resolved to general satisfaction."

CHAPTER FIFTEEN

TOUMAÏ

Is not the midnight like Central Africa to most of us? Are we not tempted to explore it,—to penetrate to the shores of its Lake Tchad, and discover the source of its Nile, perchance the Mountains of the Moon? Who knows what fertility and beauty, moral and natural, are to be found? In the Mountains of the Moon, in the Central Africa of the night, there is where all Niles have their hidden heads.

HENRY DAVID THOREAU, "Night and Moonlight" (1863)

When a Chadian undergraduate named Ahounta Djimdoumalbaye was on his first mission searching for fossils in the Djurab Desert in January 1999, Michel Brunet made a prediction. He said, in French, "Ahounta, it is you who will find it. If there is a primate here, I am sure that it is you who will discover it." Djimdoumalbaye, who was studying natural sciences at the University of N'Djamena and new to fieldwork, thought Brunet was teasing—he did not really believe him. But when Brunet had to be evacuated by car to N'Djamena because he had dysentery, he repeated what he had said to Djimdoumalbaye: "It is you who will find it."

Djimdoumalbaye would remember Brunet's prophecy a year and a half later on the last morning of a ten-day expedition to survey a sandstone basin in the western Djurab Desert of northern Chad, a fossil quarry called Toros-Menalla that had been discovered by Yves Coppens in the 1960s. Brunet and his team had found many fossils of animals there since January 1997. It was far west of the Koro-Toro fossil beds where Brunet had found the 3.5-million-year-old jawbone of Abel a few years earlier. Djimdoumalbaye was one of just four people on this short mission, along with the French geographer Alain Beauvilain, who had initially invited Brunet to Chad for a conference in N'Djamena, and two other Chadians. All were members of the Mission Paléoanthropologique Franco-Tchadienne (MPFT), a coalition of sixty researchers in ten nations formed by Brunet in 1994. In any long-term paleoanthropological mission of this size and scope, the scientific leader does not participate in every survey, and Beauvilain led this one without Brunet.

On this morning—July 19, 2001—the small team rose at seven A.M., had coffee, and parked their two pickup trucks on top of a sand dune so they could find them easily if they had to evacuate quickly. They sensed that a sandstorm was approaching, and they knew it took only a few moments for the visibility to get so bad that they could barely see one another. As usual, the four split up into pairs, and fanned out to collect fossils scattered across the basin floor. They were feeling tired and a bit morose because it was the end of the expedition. A mist was already descending on them at about eight A.M. when Djimdoumalbaye saw a ball partly covered in black manganese on the ground, about ten steps away from the edge of a sand dune. He walked toward it, and as he got closer, he saw two rows of teeth. He wondered if it was the jawbone of an extinct pig. He gently removed it from the ground, where it adhered to a layer of sandstone. He turned it over and was surprised to see an apelike face, with two vacant eye sockets, a nasal opening, and teeth. It was partly covered in black crust that looked like a cap of black hair on top. Its face was lopsided, skewed to the left, as if it had been wedged beneath a heavy weight. It was also missing its lower jaw, but, amazingly, one from another individual was found later in two pieces nearby. Otherwise, it was remarkably complete, with almost all of its teeth present and well preserved but a strange shade of red from the iron salts they had absorbed. Djimdoumalbaye had an inkling it was a long-hidden skull of a hominid from the ancient shores of Mega-Lake Chad, but he tried to resist getting too excited until a paleontologist could see it and verify its identity as a human ancestor. "I was thus alone with the cranium," he wrote later. He studied the skull, lost in his thoughts about what it could be. "Then I returned to reality and came back to earth and shouted to my Chadian colleague Fanoné Gongdibé: We have what we seek! We are victorious!"

Fanoné Gongdibé, who had been trained by Brunet in Poitiers to make casts of fossils, took a look at the skull and initially thought it belonged to a big ape. But when he saw the molars, he recognized that they might be more human than ape. He and Djimdoumalbaye gestured to Beauvilain to come toward them, telling him to get his camera, that they had found the complete skull of a big ape, maybe an ancestor of humans. Beauvilain thought it was a joke. But when he saw the skull, it was obvious that it was a major discovery. He took photos and video and used a handheld GPS device to record its precise location. The four men spent the morning collecting fossils of animals found alongside the skull and carrying them to their trucks.

Before long, they recognized some of the bones they were holding in their arms as extinct species of animals that were very ancient. Some had been alive in Africa about 6 million years ago or so—including an extinct species of elephant and an anthracotherid, an extinct family of hippopotamus-like mammals. By the time the four had collected 141 fossils near the skull, they realized that if the skull really belonged to a hominid that had lived more than 6 million years ago, it was probably the most exceptional one ever found.

The team members began to feel the full weight of the discovery. They knew that the skull was too important to be left in their hands, and would soon be sent to Brunet for study in France. Beauvilain used a satellite phone to notify Baba El-hadj Mallah, the director of the Centre National d'Appui à Recherche (CNAR), the Chadian government agency responsible for promoting research in Chad—and the person who had ordered this particular mission and signed the research permit. After a few seconds on the phone, Mallah understood the importance of the discovery. He told Beauvilain that he would alert the appropriate authorities, including governmental officials who would want to see the fossil before it was exported for scientific study. But neither Beauvilain nor Mallah called Brunet, which would have been standard procedure on other large teams. Beauvilain recalled later that one of the Chadian team members said then, "This is too big for us! It is going to bring us trouble."

~~~~~

Michel Brunet was working in his laboratory at the University of Poitiers on the morning of July 24, 2001, five days after the discovery of the skull, when he got a telephone call from CNAR director Mallah, who told him about the skull as Beauvilain listened nearby. Mallah said that the team members thought it looked like the skull of a hominid, but that it was so primitive, they were not sure. The media had been calling to ask them about the discovery, and they wanted Brunet to come quickly to N'Djamena to determine whether it was an ancestor of apes or humans.

By the time Brunet was able to get to N'Djamena almost a month later, on August 23, a long line of dignitaries including the prime minister of Chad, members of the Supreme Court, and the French ambassador had all had viewings of the skull. The French ambassador Jacques Courbin saw it during a morning meeting at CNAR and told the French newspaper *Le Figaro:* "It is very moving to hold that skull in your hands. It is a bit like the Grail for a researcher."

Even members of the media in N'Djamena had seen it. Beauvilain had drafted a short press release, which had been sent to the Chadian media the day before Mallah called Brunet. The release included three paragraphs that said that a team of four from CNAR had discovered the "well-preserved head of a fossil hominid" and 141 fossils near the skull, some of which helped the team date the site to about 6 million years. It noted that this was the time period when the human family emerged.

The local Chadian papers published the press release on July 24, although they were not allowed to take photos of the skull or to film it. A reporter for *Le Figaro*'s weekly magazine who had been in Chad in June got the press release and ran a story on its front page on July 31, with the headline A NEW ANCESTOR OF MAN? Within a few days, newspaper reporters and television and radio crews from all over the world knew about the discovery, and were badgering Brunet for comment. He was not at all pleased. He did not want to comment on fossils he had not seen. He was concerned that the viewings of the skull would jeopardize his chances for publishing a description in a scientific journal, such as *Nature,* which rejects reports on fossils that have been described in detail elsewhere. He told one French newspaper: "What if it is a pig's skull?" Indeed, Brunet remembered an earlier leak to the media about a discovery in the field of "hominid" teeth that turned out to be pig teeth. Brunet would reveal the full brunt of his displeasure about the premature announcement of the fossils to Beauvilain on August 23, en route to Chad, when by coincidence they had both booked seats on the same flight from Paris to N'Djamena. (Beauvilain, who had lived in Chad since 1989, had left for vacation in France soon after he delivered the skull in N'Djamena, and he was returning that night to Chad.) They sat beside each other, mostly in silence.

The tension between Brunet and Beauvilain vividly illustrated the difficulties of organizing and leading an international collaboration of researchers, support staff, and fossil hunters from afar. Brunet knew well that without his team he was "nothing," as he would often say, and that it was important to build a certain esprit de corps to be successful in such daunting conditions. As team leader, he was responsible for the safety and the well-being of the team members—and they had to be a cohesive unit as they headed into the furnace of the Djurab, where they faced illness, vehicles that broke down, sandstorms that lasted for days, scorpions, snakes, land mines, and tedious work. Like other team leaders, Brunet raised most of the money and selected many of the researchers and graduate students to join the Franco-Chadian mission. But, in return, he needed lieutenants in the field he could count on through thick and thin—people who would be team players loyal to him in the face of crisis and triumph, when the lure of instant fame can splinter a team. As a result, he was deeply wary of any attempts to undermine his authority or goals for the mission, or to leak discoveries to the press prematurely.

This was perhaps the first time that a major discovery was announced to the press and shown to high-level government officials before the scientific leader of the team had seen the fossil. Brunet knew that Beauvilain was invaluable at organizing the surveys for fossils and arranging logistics in Chad; indeed, Beauvilain had gotten to know the desert well in twenty-eight scientific missions with Brunet and with geological surveys over the years. But Beauvilain was highly independent and answered to Chadian and French authorities as much as to Brunet—and they did not always see eye to eye on how to run operations in the field or how to manage the growing fossil collection in N'Djamena. Brunet demanded loyalty from the team members, especially in a political climate where no one's claim to a fossil site could be taken for granted. Beauvilain would protest later that as a geographer, he never had any intent of taking Brunet's place as scientific leader but that he wanted more credit for his role in the mission and for the part he and several Chadians played in the discovery of Toumaï. Regardless, the episode marked the beginning of the end of Beauvilain's collaboration with Brunet.

Brunet finally saw the skull a few days after arriving. He described the moment a year later: "It's a lot of emotion to have in my hand the beginning of the human lineage. I have been looking for this for so long, I knew I would one day find it, so it is a large part of my life too. I've been looking for twenty-five years." Brunet also told the media he was not surprised that Djimdoumalbaye had found it because he was the best fossil hunter on the team.

On August 31, Brunet and Beauvilain showed the skull to Chadian president Idriss Déby. Déby spent two hours with them, expressing a deep interest in paleontological work and the areas of the Djurab Desert where they had searched for fossils; he knew the desert well, because he had lived there for many years. Déby kept a delegation waiting for two hours at the airport to go on an official trip to Libya as he examined the skull and talked with Brunet. Before the audience with the skull was over, Brunet suggested that President Déby give a two-syllable name for this ancient member of the human family. After a minute of reflection, President Déby proposed the name Toumaï. He explained that in the Goran language it was a name given to children who were born during the hot, dry season in the desert, when it was most treacherous to be a newborn—thus, Toumaï means "hope of life," and is supposed to bring babies luck, allowing them to live through difficult times. It was a fitting name for the fossil with great potential, whose offspring would have had to survive difficult times to have given rise to humans.

Just before midnight the next night, Chadian officials gave Brunet the skull to take with him back to France for a short time for scientific study. The next morning, on September 2, 2001, Brunet boarded a flight for Paris with Toumaï, the 7-million-year-old skull that would travel more after its death than during its life.

Once Brunet had the skull in Poitiers, he cleaned it up, made casts of it, and began to contact leading researchers to help him in the intensive analysis of such an important specimen. Before Brunet discovered Abel in 1995, his specialty was ancient mammals, not hominids. He had learned a lot about hominids since then, but he knew that the analysis of a skull would require special skills, ranging from understanding the microstructure of the teeth to using CT scans to reconstruct a cast of the skull. Brunet would seek the advice of some of the world's best experts to discuss his analysis of this skull— a pattern he repeated over the years, pulling together an impressive brain trust to collaborate with him on Toumaï and the ancient animals found with it.

He would take Toumaï on a world tour, heading first to Harvard University to see his longtime friend David Pilbeam. It was right after the attacks on the World Trade Center on September 11, 2001, and the sight of so many American flags on cars and in windows moved him. He would keep a small American flag in his office as a reminder of the solidarity among Americans at that time. Once he was at Harvard's Peabody Museum, showing Pilbeam the skull of Toumaï, the hard work—and the fun—began. Pilbeam invited a former student of his who had just joined the faculty of anthropology, Dan Lieberman, to meet Brunet. Pilbeam introduced Lieberman to Brunet, and then, with a smile on his face, showed him the skull without saying a word about it. "As I looked it over, I went over its features in my mind, looked up, and said, 'Holy shit!' " recalled Lieberman.

What surprised everyone who saw the skull was its mix of ancient and more derived traits—what paleoanthropologists invariably call a "mosaic" of apelike and more human features. It had a brain the size of a chimpanzee's and widely spaced eyes like a gorilla's. Yet from the front it had a short and flat lower face with derived traits found in later hominids. Its mug did not jut out as much as a chimpanzee's, and it had a massive brow ridge, which suggested that it was a male. But Toumaï's upper canine was too small and unsharpened to belong to a male ape, and both its upper and lower canines were worn on the tips, which suggested they were not used for cutting as in an ape. It was also the most complete skull of an ancient hominid ever found—95 percent of the skull was still there, minus the lower jaw.

Pilbeam agreed with Brunet that Toumaï was a hominid, even though it would be best to have limb bones to prove that it had walked upright—still a traditional marker of being a hominid. But if it was a hominid, what type? From Harvard, Brunet took Toumaï to Tempe, Arizona, to compare it with the best skull of Lucy's species, which was being studied by paleoanthropologist Bill Kimbel at the Institute of Human Origins at Arizona State University. He then headed to Berkeley, where he met with Tim White and Clark Howell, and they put the cast of Toumaï alongside casts from their remarkably complete collection of replicas of fossils of early human ancestors. White noted that the skull was pretty distorted and sandblasted, but he recognized traits that it shared with *Ardipithecus ramidus,* particularly in the canines and other teeth. He thought it possible that Toumaï could be a member of the same genus as *Ardipithecus,* but it was about 2 million years older and clearly showed differences that put it in a new species, at the very least. White agreed that it appeared to be the earliest hominid cranium known, although he would add the caveat that it was technically a "dental" hominid, since there were no skeletal parts to prove that it had walked upright. "This is by no stretch of the imagination a chimpanzee," he said.

After his American tour, Brunet traveled to Ethiopia and Kenya to compare the skull with the original fossils of other early hominids. He took Toumaï to Zurich to get its head examined by a pair of researchers who used industrial CT scans to see inside the skull and to reconstruct it. Back in Paris,

Brunet brought Toumaï to the National Museum of Natural History, where he met Pickford and Senut in their basement laboratory. Brunet showed them the new skull from Chad. Pickford and Senut generously left Brunet alone with the fossils of *Orrorin*, so he could compare its teeth directly with those of Toumaï. It would be one of their last amiable meetings.

Brunet and his colleagues eventually concluded that Toumaï was a new kind of hominid that belonged in a genus of its own. But what to use for a formal name? On a trip back to Chad, he met with President Déby and other Chadian authorities who urged him to use the name Sahel, to note the skull's origins in the sub-Saharan region of the Sahel of Chad. Brunet agreed to use both names, and he proposed the name *Sahelanthropus tchadensis*, drawing on the Greek root *anthropos* for "human being." The name suggested that the skull and teeth belonged to a Chadian species of a new genus of human beings from the Sahel. Brunet also returned to the desert, where he went to the Toros-Menalla fossil beds to see where the skull was found. While he was there, on January 20, 2002, the same skilled fossil hunter, Djimdoumalbaye, found another tooth and another jaw fragment from the same type of hominid as Toumaï.

Almost a year after its discovery, the skull of Toumaï appeared on the cover of *Nature* on July 11, 2002, under the headline THE EARLIEST KNOWN HOMINID. The remarkably complete skull appeared as if it were floating in space, high above the Djurab Desert, hovering like a bodiless apparition over the paleontologists' tents in the sand below. In the first of two reports, Brunet and thirty-seven coauthors described the skull as the oldest and most primitive member of the human family—a hominid so ancient that Brunet said that Toumaï could "touch with its finger" the last ancestor shared by humans and chimpanzees.

If Brunet and his coauthors were right about Toumaï being an early human ancestor, it proved that hominids had already spread far and wide in Africa soon after they arose. Brunet emphasized that the skull was found twenty-five hundred kilometers west of where the 6-million-year-old fossils of *Orrorin* and *Ardipithecus* had been discovered in the Great Rift Valley, once considered the cradle of humankind. With some glee, Brunet drove a stake through the heart of the now-moribund East Side Story proposed by his long-time friend Yves Coppens, in which humans and their ancestors arose on the east side of the rift and the African apes evolved on the west side.

Brunet calculated Toumaï's age by the company it kept. The rich collection of seven hundred animal fossils from the Toros-Menalla fossil beds indicated that Toumaï had been alive sometime between 6 million and 7 million years ago. Some of the species of animals, including an extinct species of pig, were the same species as those found at two well-dated sites in Kenya and Ethiopia. Until recently, there were no volcanic tuffs at Toros-Menalla, which researchers could have used to date the sediments with more precise radiometric methods, so the team had to rely on the evolution of fossils of mammals to date the skull.

The two papers announcing the discovery had the impact of a "small nuclear bomb," said Lieberman. Indeed, the seismic waves from the announcement were felt around the world. It was the third fossil in three years to push back the age of the earliest hominids—now to 6 million to 7 million years ago. Paleoanthropologist Bernard Wood, who had moved to George Washington University, in Washington, D.C., wrote the accompanying commentary in *Nature*, as he had done for *Ardipithecus ramidus*. "A single fossil can fundamentally change the way we reconstruct the tree of life," Wood wrote. "Even if it is a fossil ape not directly related to humans it is still the first glimpse we have had 6 million to 7 million years ago."

Three months after Toumaï appeared on the cover of *Nature*, Brunet was sitting at his computer reading his e-mail in his office at the University of Poitiers, when he opened a message from Milford Wolpoff, a paleoanthropologist at the University of Michigan in Ann Arbor whose specialty is Neandertal and modern human origins. Wolpoff wrote that he would be in France soon, and would like to see the skull of Toumaï. Brunet was irritated. Wolpoff and a former student of his, John Hawks of the University of Wisconsin,

Madison, had joined Martin Pickford and Brigitte Senut to write a letter to *Nature* challenging Brunet's description of Toumaï as a hominid. Senut had told reporters from the day the *Nature* paper appeared that she thought Toumaï was the female ancestor of gorillas. Now the four had written a letter that was in press in *Nature*. It would be published the following week, on October 10, 2002, under the title *"Sahelanthropus* or *Sahelpithecus?"* The name *Sahelpithecus* means "ape from the Sahel" as opposed to "human being from the Sahel."

In other words, the four proposed that Toumaï was an ape. They wrote that the features in the teeth, face, and skull used to classify Toumaï as a hominid were not unique to hominids—and could be found in ancient apes. For example, the canine might be small and unsharpened because Toumaï was a female—not a male hominid with a reduced canine. They also measured the length and angle of a plane at the back of the skull where the neck muscles had once attached (based on the photographs in *Nature,* not the skull) and concluded that it resembled the angle in apes, not hominids. This was significant because it challenged the one bit of indirect evidence that Toumaï might have walked upright.

For Brunet, the letter was deeply annoying. In what should have been a moment of triumph, he was facing headlines such as the one in the *Observer* of London that said, FOSSIL FIND OF THE CENTURY MAY JUST BE A GORILLA AFTER ALL. Brunet's reply in *Nature* began with the observation that the Taung baby had also been described as a gorilla when Dart first proposed it as a hominid in 1925—and that any human ancestor alive this close to the split with chimpanzees would, indeed, have many primitive traits. Brunet then pointed out that Wolpoff and his fellow letter writers had measured the angle at the base of the skull incorrectly, failing to take into account that the skull was distorted. The correct angle, minus the effects of being squashed under the weight of sand and rock, put the angle and length within the range expected for early human ancestors, not apes that walked on all fours. Furthermore, he defended the original analysis, saying that the all-important upper canine was diagnostic of being a hominid, because its size and proportion to other teeth fit with human ancestors better than with apes. With a tone of indignation he wrote, "These authors not only misrepresent the specimen's morphology, but also fail to identify a single character to support their suggestion that Toumaï is a gorilla rather than a hominid ancestor." Instead, he suggested that they were ignoring the evidence that Toumaï was a hominid in favor of their own "belief" that *Orrorin* was a hominid. Wolpoff, Pickford, and Senut would all protest, saying that they were commenting on a published description of a fossil—and that it was the process of science to be able to debate and interpret the fossil evidence.

But at his desk in Poitiers the week before the letter was published, Brunet made it clear that he considered it a personal attack rather than a scholarly debate, because none of the four authors had seen the fossils in any detail before they wrote their letter. They had gone on the attack in *Nature*, rather than corresponding with him about particular details. He took particular umbrage at the flippant title, waving a cast of Toumaï in one hand and the skull of a gorilla in the other. "It is absolutely not a paleo-gorilla," he fumed. If so, it would have been the first gorilla that lived on the edge of a desert, he added. Needless to say, Brunet did not invite Wolpoff to Poitiers to see the fossils.

Only three months after the publication of Toumaï in *Nature*, Brunet was besieged by requests from colleagues and journalists who wanted to talk to him or see the fossils. Although he had toured more labs in the course of a year than any other discoverer of a hominid fossil in recent memory and included enough coauthors on the Toumaï manuscript to please the editors of a medical journal (where papers are often published by large teams), he was being asked to share the skull even more widely. At the same time, journalists from New York to Tokyo had been coming through his lab regularly to interview him. A producer from the Discovery Channel had taken him to lunch, and a French film crew was planning a documentary on the discovery of Toumaï. *National Geographic* had offered him a lucrative contract as an explorer in residence in exchange for exclusive rights to his story and future research findings (Brunet declined the offer, noting that "Americans want to buy everything"). At age sixty-two in the fall of 2002, with a heart condition, he felt worn down and resentful of the demands on his time. "It's a tax," Brunet said, with deep ambivalence about his instant fame. "Sometimes it is terrible. I am so tired I don't want to speak. I have decided I am doing this to the end of October. Then I stop. I am tired."

The worst was yet to come. The following September, when Brunet hosted the president of Chad, Idriss Déby, in Poitiers as part of Déby's official trip to France, the French newspapers were full of embarrassing reports of a "paternity battle" over Toumaï. The French geographer Beauvilain, who had led the four-man search party that found Toumaï, had published a book in which he described the discovery in detail. He also complained that Brunet had failed to give him proper credit for his role in the discovery. Brunet had listed him as coauthor on the manuscripts in *Nature* for Toumaï and the jawbone of Abel, which was more credit than most team leaders give team members who are not scientists; support staff, fossil hunters, and logistics organizers usually get acknowledged at the end of a report. But the press releases had described the Chadian student Ahounta Djimdoumalbaye as the sole discoverer, and Beauvilain thought that he and the other two Chadians deserved more credit. Then, on the day that Toumaï was published, the French foreign service had recalled Beauvilain to France. He had moved back unwillingly, uprooting his Moroccan wife and their children after fourteen years in Africa. He got a job as a teacher at the University of Paris X–Nanterre, but he was still angry with Brunet when he published his book nine months later. The collaboration between Brunet and Beauvilain was over.

Nonetheless, Brunet readily admits that there is a positive side to the intense scrutiny and attention. In October 2002, Brunet was awarded a medal of honor from the Council of the Poitou-Charentes Region, with a reception in his honor attended by politicians and scientists in charge of research and funding in his field in Paris. Brunet joked after the reception that he needed funds to buy a new vehicle for his expeditions more than medals, but that would come soon. The following year he was awarded $1 million from the Israel-based Dan David Foundation, which gives out awards for impressive discoveries. He also was selected for France's highest honor—to become a knight of the French Legion of Honor. Soon after, he had a private audience with French president Jacques Chirac.

Still, Brunet did not forget where he came from. On the same day that he got the medal of honor from the regional council of Poitou-Charentes in Poitiers, he drove to the small farming town of Neuville de Poitou, where a university colleague was the mayor. In the town library, he gave a slide show and lecture to a standing-room-only audience of about seventy-five men, women, and children, many of whom seemed well versed on human evolution. This was a sharp contrast to the United States, where it would be hard to imagine a crowd in a small rural town turning out for a scientist's talk on human evolution—indeed, 45 percent of Americans who participated in a Gallup poll in 2004 thought that "God created human beings pretty much in their present form about 10,000 years ago." Brunet can be a dramatic speaker, and the audience stayed past eleven P.M. while he answered every last question. By the time the mayor finally stood up to thank him, Brunet was visibly exhausted. But he also was philosophical. "It's part of what you have to do," he said. "We are scientists and part of our job is to communicate. But for me, the best time shall be to work with the fossils."

Wisdom

of the

Bones

The question of questions for mankind—the problem which underlies all others, and is more deeply interesting than any other—is the ascertainment of the place which Man occupies in nature and of his relations to the universe of things.

THOMAS HENRY HUXLEY, 1863

CHAPTER SIXTEEN

BONES OF CONTENTION

Arrange whatever pieces come your way.

VIRGINIA WOOLF

The French Academy of Sciences has been the scene of many passionate, even dangerous, debates since its formation in a Jesuit monk's cell in 1666. The founding members met to discuss the radical ideas of Galileo and Descartes. A century later, in 1793, during the Reign of Terror, the elitist academy was deemed undemocratic and abolished; the renowned chemist Antoine Lavoisier was beheaded for his prominent role. But the academy reconvened two years later and moved into its present home in the magnificent baroque palace of the Institute of France on the Left Bank of the Seine River. The academicians were still perceived as a political threat, with Napoleon complaining about the "salon politics of liberal intellectuals." He appointed himself president of the institute in 1801, the better to reform the academies to his liking. The Academy of Sciences survived, and by the late nineteenth century, the fraternity of male scholars would move to protect itself from a new type of threat: it would vote to reject Marie Curie for membership in 1911, only months before she won her second Nobel Prize. Even the shroud of Turin was brought here—with scientists arguing intensely for two days over whether it was legitimate and worthy of study.

It was thus a perfect setting for staging a passionate debate among three scientists whose fossils were the trio of leading contenders for the status of the oldest known hominid. On a gray and drizzly Monday in September 2004, Michel Brunet, Brigitte Senut, and Tim White met for a rare face-to-face debate, along with French paleoanthropologist Yves Coppens, a member of the academy. Word of this unprecedented lineup had spread widely, and paleontologists, archaeologists, and anthropologists from all over Europe jammed into the opulent Grande Salle des Séances, where the speakers were poised to channel humanity's most ancient ancestors.

Translators took their places in a soundproof glass booth and headsets were passed out to members of the academy, so they could listen to the panel discuss the topic of "The First Hominids" in English or French. Television crews stood by, and the heat of their lights made the dark, wood-paneled room even more sweltering, with the windows closed and the draperies drawn for PowerPoint presentations. As Brunet, Senut, and White took their places in turn on the raised dais to speak, the camera lights illuminated the portraits and marble busts of the author Voltaire, the philosopher Jean-Jacques Rousseau, the mathematician Pierre de Fermat, and the physicist Charles-Augustin de Coulomb. Perhaps even more telling, one gilt frame was empty, a reminder to the French scientists that one among them might eventually have his or her visage framed there as well.

The first one to speak was Michel Brunet. Dressed in a navy jacket, striped shirt, and tie, with his straight gray hair brushed back, he was ready for serious business. He quickly dispensed with niceties to get right to the "bone of the matter," in his words. He held up a fossil jaw and said, "Last month, this jaw became part of history." The bone in his hand was a partial lower jaw of Toumaï's species, *Sahelanthropus tchadensis*. It had been the subject of an article in a recent issue of the *South African Journal of Science*. In that article, the French geographer Alain Beauvilain, who had been part of the team that found the jaw, charged that Brunet's team had glued an isolated molar—a wisdom tooth—on the wrong side of a lower jaw. Beauvilain and a French orthodontist who was his coauthor, Yves Le Guellec, said the isolated molar was glued on the right side instead of the left side of the jaw, where they thought it belonged. It was a curious report; neither one was a paleontologist or an anatomist familiar with fossil teeth, and their criticism did not alter the fossils' status as a new type of human ancestor in any way. But it cast a cloud over Brunet's methods, suggesting that his team was sloppy with their fossil analysis and inventory methods. The spectacle of scientists fighting tooth and nail over a wisdom tooth was widely reported in the French media.

Brunet brought up the report at the beginning of his talk at the academy, complaining that he had first learned of the article when he was visiting White in Berkeley, after it was published. The journal had not given him a chance to respond before publication, so he would do so at the meeting. He held up the jaw and explained that the wisdom tooth was found ten centimeters beside it, it was cleaned, and it was glued in where he said there was an "unambiguous match" between the molar and roots on the right side of the jawbone, which was evident on CT scans of the jaw. After he took a hard look at Senut, he said that a certain paleontologist who had not seen the jawbone had been quoted in the French press as saying he thought the tooth was glued on the wrong side. "I have to admit, I wonder what his intentions are," Brunet said. Then he mentioned that this still unnamed paleontologist—assumed to be Senut's partner, Martin Pickford—was one of the codiscoverers of *Orrorin*. Pickford, in fact, had translated Beauvilain's article into English for the journal. Brunet's intent was clear: he believed that Pickford was waging a campaign against *Sahelanthropus* as a hominid and against Brunet's ability to analyze the fossils, because it threatened *Orrorin*'s status as the earliest member of the human family (a charge that Pickford said is "absurd"). Brunet then went over the traits that made Toumaï a hominid and said, "Until proven otherwise, *Sahelanthropus tchadensis* is the oldest known hominid."

Pickford, who was traveling and could not attend the meeting, would say later that the editor of the journal had asked him to translate the article, not Beauvilain. He at first wrote a letter of apology to Brunet that he sent to the South African journal, but he rescinded it after Beauvilain and Le Guellec refused the apology. Then, twenty-seven researchers signed a letter in support of Brunet, saying that the molar had been put in the right spot. Pilbeam, who was a coauthor with Brunet on the report on Toumaï, said he was stunned that the journal would print Beauvilain's article because "any competent morphologist seeing the originals or casts, along with the CT scans, would immediately recognize that the molar is from the right side." But Pickford and Beauvilain criticized the letter as a tactic to intimidate and stifle debate on published fossils, and Beauvilain wrote a new letter to the journal with more criticism of Brunet's methods.

Senut, who looked uncomfortable during Brunet's comments, did not respond. She looked particularly unhappy when a member of the audience who'd been one of the twenty-seven to sign the letter in support of Brunet—human paleontologist Marie-Antoinette de Lumley of the National Museum of Natural History in Paris—spoke up after the panel discussion to defend Brunet. She said, "It is obvious that Brunet was correct—this is a right-sided molar that corresponded to the right-sided jaw." De Lumley then scolded those behind the journal report for taking such a negative approach; if a mistake had been made, she said, there were better ways to handle it than exposing it in the media.

Senut was up next, but with evident self-restraint she stuck to the scientific topic at hand—namely, the evidence for *Orrorin*'s status as upright-walking member of the human family that still lived in the trees. Dressed in a simple gray skirt and floral blouse, she chose her words carefully. At the end, however, she lobbed a shot at White. "I hope that Tim White will tell us about *Ardipithecus ramidus*," she said, reminding everyone that they had been waiting a decade for details on the unpublished skeleton.

White took the microphone—and the challenge. Dressed neatly in a tan jacket, tie, and slacks, he also had some unfinished business of his own to attend to. After a quick overview on how to analyze a thighbone to tell if its owner walked upright like a human or in some other way, he got to Pickford and Senut's research. He began to slice and dice the CT scans of the thighbone of *Orrorin* that Senut had just shown, pointing out that they had been published twice—and that their poor resolution did not improve with the second publication, even if it was in the reputable journal *Science*. He agreed that the external features showed that *Orrorin* could walk upright, but he said the central question in a hominid this old was, How did it walk upright? Did it walk upright like a human, as Senut and Pickford claimed, or like an australopithecine—or in some novel manner?

He reminded the audience that Pickford and Senut had proposed that *Orrorin* walked upright with a more modern gait than Lucy's species and other australopithecines, therefore bumping them off the direct line to humans. But he complained that Pickford and Senut still had not presented the obvious evidence to support that claim—in the form of photos, X-rays, or reliable CT scans that showed the distribution of bone density inside the neck of the thighbone. He was intent on getting the inside story on *Orrorin*'s thighbone, because there was no other thighbone this old—not for *Sahelanthropus* or for *Ardipithecus*, he said. After White made this point, he stopped his presentation to ask Senut directly if the thighbone had been cracked or broken through the neck when they found it.

She answered, "It was broken."

White then posed a series of questions: Why didn't they measure the distribution of bone thickness directly before they glued it together? Why didn't they provide a photo? Why take CT scans when they could have a direct photo, which would be better?

Clearly put on the defensive, Senut shook her head from side to side and said that she had measured the thickness of the bone and that the distribution fit with what she expected in a human thighbone. Later, Pickford added that it was difficult to photograph the inside of the broken bone to provide conclusive evidence because of the zigzag pattern of the break.

White retorted that without reliable data, their hypothesis that *Orrorin* walked like a modern human was extraordinary. "There's a creationist position here—*une position créationniste!*"

Senut protested indignantly: "I am not a creationist! Otherwise I would not study evolution for so many years!"

In the middle of the heated exchange, Senut reminded him of her question about the status of *Ardipithecus ramidus*. White responded by showing images of the crushed skull (the first time he had done so for anyone outside of his team), commenting that this was the "most fragile skeleton ever found." In a humorous moment that deflated some of the tension in the room, he defined the term "roadkill" for the French audience.

He showed a computer movie composed of micro-CT scans taken of the skull. The CT scans were startling: the top of the vault of the skull had been crushed to within an inch of the base of the skull, forming a one-inch-thick slab of hundreds of small pieces. It revealed the tremendous task he faced to reconstruct it—a task that involved micro-CT scans of each piece, which White and his colleagues were using to help reconstruct the skull and see its internal structure. He said, "I'm sorry it's taken us so long to analyze this fossil, but I think you want the right answer instead of the quick answer."

Finally, as tempers flared in the sweltering room, the French science journalist Sophie Coisne of *Science et Vie* stood up at the microphone during the question-and-answer session and asked the panel, "Why do you scientists always argue about your fossils? Why don't you share the fossils?" White was the first to respond, saying indignantly, "I take personal offense at that comment." Then he explained that there was a process in science called peer review. Brunet's paper in *Nature* describing the fossils of *Sahelanthropus* as a hominid had been subjected to six reviews by anonymous scholars before publication, which was intense scrutiny of a manuscript. For Beauvilain, with a translation by Pickford, to "recklessly" accuse Brunet of placing the molar in the wrong side was irresponsible. Yet Pickford and Senut would not provide their peers with a reliable image of the interior of the thighbone. At that point, he gestured to the portraits and busts of famous French scholars on the walls and said, "Even these gentlemen on the walls understand the peer-review process. This is not about science. This is about theater. This is theater!"

⋙

If Sophie Coisne's exhortation to share their fossils could work, and if all the teams could overcome their differences to get together with their fossils in a sort of Yalta summit of paleoanthropology, what would they bring to the table? Of course, this is easier said than done. Even in the unlikely event that they all agreed to share original fossils, rather than casts or three-dimensional computer images, they would have to get special permission to remove the fossils, because the specimens are the priceless property of the nations where they were found. But for the sake of this thought experiment, if the researchers were able to gather every single tooth, jaw fragment, and finger or toe bone they had found that was 4 million years or older—and that had been published between 1994 and 2005—they would bring 144 fossils from Chad, Ethiopia, and Kenya to the summit.

If they were really in a munificent mood and brought along unpublished fossils, the number would soar to include the bones of two partial skeletons from Ethiopia: the crushed partial skeleton of *Ardipithecus ramidus* being analyzed by White and his colleagues and a new partial skeleton of an upright-walking hominid that was nearly 4 million years old and had been discovered in February 2005 by Yohannes Haile-Selassie in Ethiopia. Two other teams working in Ethiopia, including those of Ethiopian Sileshi Semaw and Austrian Horst Seidler, have also found new fossils of *Ardipithecus* that should fill in some gaps for that species.

Even though most of the published fossils are teeth and jaw fragments, it is an impressive collection by any standard. Before 1992, only a half dozen fossils of hominids older than 4 million years were known. Together, the fossils collected in the 1990s and early 2000s would cover a large desk and would represent a few dozen individuals at least. This is the hard data (along with younger fossils) that the field of paleoanthropology is built upon. In a proposal for funding to the National Science Foundation in 2003, Berkeley paleoanthropologist Clark Howell and White wrote, "These dramatic discoveries promise to finally answer questions that have been a central focus of human origins studies for over a century: *when, where, how,* and *why* hominids originated from the last common ancestor they shared with chimpanzees."

Although paleoanthropologists are asking many questions of these fossils, they are just beginning to get answers about their place in nature. The *when* and *where* are clearly coming into focus: *when* was more than 6 million years ago, possibly 7 or 8 million years if the oldest fossils on the table prove to be the earliest members of the human family. This means that paleoanthropologists can now see back in time twice as far as they could before the discoveries of the past decade. The new fossils have also meant that the fossil record is finally in sync with the molecular clock, which for thirty years has timed the split of chimpanzees and humans to sometime in the past 4 million to 6 million years—a range revised recently to 5 million to 7 million years.

The *where* is clearly Africa, which has been widely accepted as the birthplace of the human family since the discovery of Lucy's species in the mid-1970s. But what is new is that the fossil record is exclusively African for almost five million years, since no hominids older than 1.8 million years have been detected outside of Africa. These new fossils at the dawn of humanity are a dramatic confirmation of Darwin's tentative proposal in 1871 that "it is somewhat more probable that our early progenitors lived on the African continent than elsewhere."

But *who* among the fossils on the table is the oldest known progenitor of humans? If Brunet were to bring to the fossil summit the stunning skull of Toumaï and the jawbones and teeth of other members of *Sahelanthropus tchadensis* that were alive 6 million to 7 million years ago in Chad, would everyone assembled agree that it was the earliest known hominid, as he and his many coauthors have asserted?

The problem is that there is still no consensus on how to define a member of the human family this ancient. There are no fossils of chimpanzees or gorillas—other than some isolated teeth—to show paleoanthropologists what the ancestors of the African apes looked like 5 million to 7 million years ago. Even though Toumaï and all of the other fossils on the table have been proposed as hominids, they were alive so close to the split with chimpanzees that they look more like the ancient ape ancestor they shared with chimpanzees than their human descendants. As White told the well-known radio talk-show host Terry Gross on National Public Radio's *Fresh Air:* "You wouldn't invite *Ardipithecus ramidus* to dinner."

Until the mid-1990s, the litmus test for being a member of the human family was walking upright—the first major shift in anatomy after the split from apes, before the brain expanded and our ancestors began making tools and speaking in languages. None of the living or known extinct apes walked upright. This test, therefore, worked to identify hominids that were 4 million years or younger, because by then all of the australopithecines and members of the genus *Homo* were running on two legs.

Unfortunately, there are still too few published bones from the bodies of the earliest fossils to ascertain how they moved. The oldest fossils are so ancient that they may not have walked upright like australopithecines or humans. There may have been an awkward period of transition when they were moving from walking on all fours to two legs, and the signature of their gait on their bones may be different from the signs of upright walking that come later. Therefore, for practical reasons, a number of researchers have been identifying the first members of the human family on the basis of subtle evolutionary novelties in the teeth and skull that set the earliest fossils apart from apes. By those criteria, Toumaï meets the test for being the earliest known member of the human family.

Pickford and Senut can bring to the table a 6-million-year-old thighbone that shows signs of belonging to a creature that already walked upright. If the Millennium man, *Orrorin tugenensis,* was walking upright by 6 million years, does that make it the earliest known true hominid—even if the isolated canine teeth assigned to it look more primitive in shape than the canines of Toumaï? Are the clues from the teeth and skull more informative about hominid status, or are those from the limb bones? To prove a fossil's hominid status beyond a doubt, the definitive test is still evidence of upright walking, though more and more paleoanthropologists are debating whether a reduced canine and bigger cheek molars came before upright walking—especially if they are confident that the smaller canine belonged to a male.

Even if most of the researchers at the fossil summit could agree on which fossil was the earliest known member of the human family, they also would face the even thornier problem of figuring out how many different types of hominids were lying on that table. One reason for the confusion is that, other than teeth, there is little overlap in the skeletal parts to compare. Brunet has a skull, but no one else does, excluding White's unpublished crushed skull of *Ardipithecus ramidus.* Pickford and Senut have a thighbone of *Orrorin,* but it is not connected to a toe bone, which is what Haile-Selassie has for *Ardipithecus kadabba,* while Meave Leakey and Alan Walker have a shin-bone and a wrist bone for *Australopithecus anamensis*—and so on.

For now, they will have to stick to the teeth and jaw fragments for direct comparisons between these early hominids—at least until White and his colleagues bring a raft of brittle bones to the table when he publishes the reconstruction of the partial skeleton and squashed skull of *Ardipithecus ramidus.* White has proposed this skeleton as a Rosetta stone that will reveal the anatomic code for upright walking in early hominids. By looking at the way the torso, pelvis, and limb bones fit together, researchers should be able to see the way *Ardipithecus*'s body adapted to its movements, both on the ground and in the trees. A key question that White and others are asking of this skeleton, for example, is whether the anatomy for upright walking was already fully assembled in *Ardipithecus* at 4.4 million years, as it appears to have been in its proposed descendant, *Australopithecus anamensis,* at 4 million years. "If we had skeletons of *Sahelanthropus, Orrorin,* and *Ardipithecus,* we'd be able to see the evolution of bipedalism," says locomotion expert Owen Lovejoy, who is part of the Middle Awash team analyzing the skeleton of *Ardipithecus ramidus.*

The new fossils have clearly opened new windows into the past, in time and space, offering a view for the first time of the earliest stages of human evolution in central Africa, as well as in eastern Africa and South Africa. But when different paleoanthropologists look through those windows, they do not see the same thing. On the one hand, Tim White and Brunet see Toumaï as a close relative of *Ardipithecus,* which in turn gave rise to the australopithecines and early *Homo. Orrorin* may also be a member of *Ardipithecus,* in White's and Haile-Selassie's view. Pickford and Senut have rejected that and nominate *Orrorin,* instead, as the ancestor of humans, *Ardipithecus* as the ancestor of chimpanzees, and *Sahelanthropus* as the ancestor of gorillas or some extinct ape.

On the other hand, different researchers would look at these fossils and wonder what is missing from the table—the hominids that might have existed but have not yet been detected. Bernard Wood and others have suggested that paleontologists have just scraped the surface of the fossil record in

Africa 5 million to 7 million years ago. Toumaï's human aspect suggests to Wood that it was just one of a diverse array of hominids alive in Africa, some of which had an uncanny resemblance to hominids that came later but were not necessarily direct ancestors.

This model would see human evolution as a series of radiations of fossil ground apes that mixed and matched traits, such as upright walking and reduced canines, as they adapted to different habitats. The paleoanthropologist Ian Tattersall of the American Museum of Natural History sees the fossils of *Sahelanthropus, Orrorin,* and *Ardipithecus* as evidence of trial and error rather than a "straight-ahead slog" from primitiveness to modern humans.

This second view, if correct, would fell the tree of human life with its few strong branches. Its proponents would plant a bush in its place, full of limbs leading nowhere, with multiple species of early humans competing for prime terrain and the chance to pass on their genes to the next generations. That would make Toumaï and *Ardipithecus* and *Orrorin* not direct ancestors, like one's great-great-great-great-grandparents, but cousins of those grandparents whose own lineages died out long ago. White and others, including Alan Walker, point out that there was only one species and one genus of hominid known between 6 million and 7 million years ago. "Since when is that evidence for species diversity?" White has protested.

Paleoanthropologists may never be able to sort out how many different types of early human ancestors were on the scene in Africa in the beginning or how they are related to each other. "Is the outlook completely gloomy?" wrote Henry Gee in *Nature*. "Perhaps not. The accumulating data on paleoenvironments should at least improve our understanding of the lives and times of early hominids (and perhaps of early chimps)."

CHAPTER SEVENTEEN

HABITAT FOR HUMANITY

What is special is not the finding of hominids. This is not a treasure hunt with the hominids being the only things worth seeking. It is the knowledge of their anatomies, their behaviors, their worlds that is the "prize" here. The challenge is wresting knowledge from the very distant past by using all the resources, human and technical, that can be brought to bear on the mystery.

TIM WHITE

You can know the name of a bird in all the languages of the world, but when you're finished, you'll know absolutely nothing whatever about the bird. . . . So let's look at the bird and see what it's doing—that's what counts. I learned very early the difference between knowing the name of something and knowing something.

RICHARD FEYNMAN

At an elegant reception to commemorate the one hundredth anniversary of the birth of Louis Leakey at Chicago's posh Casino nightclub in October 2003, Michel Brunet was animatedly describing the type of environment where one ancient ape might have taken a giant step for mankind by standing up and walking upright. Brunet had recently visited the Okavango Delta in Botswana.

As he'd explored the lush floodplains in the middle of the Kalahari Desert, he had wondered if it was a modern analogue of the type of sprawling desert oasis where Toumaï had lived in ancient Chad 6 million to 7 million years ago. From the air, the six-thousand-square-mile Okavango floodplains looked like a mat of grass and papyrus reeds interwoven with freshwater channels, lagoons, and wooded tropical islands sprouting palm, date, and fig trees. On the ground, the delta was teeming with wildlife, including crocodiles, snakes, elephants, hippopotamuses, lions, cheetahs, leopards, and other modern counterparts of the animals Brunet and his colleagues had found in the ancient floodplains of Mega-Lake Chad.

At the Leakey Foundation party, Brunet mentioned that the Okavango Delta might be an apt model for the birthplace of the human lineage to a friend of his, Harvard primatologist Richard Wrangham, who had encouraged him to go to the Okavango. That's when Wrangham began to act a bit like an ape himself, showing Brunet his notion of how one ancient primate might have stood up on its hind legs to wade across a lagoon, in search of ripe fruit or mates. In their suits and ties, Wrangham and Brunet playfully demonstrated how an ape might have gingerly waded through the water, keeping its head and hands dry, though presumably without the additional requirement of holding up a glass of champagne. Enough crossings through the swamp, and some apes might have begun to adapt to this odd form of locomotion if the males could get more fruit or carry it to their mates, for example, in exchange for sex. If the upright-walking apes were better fed and produced more offspring over shorter intervals than their counterparts who walked on all fours, bipedalism eventually might have become the dominant form of locomotion. Brunet and Wrangham were, of course, only half serious about the scenario but were having a grand time at the party exploring the possibilities.

When it comes to figuring out the origins of upright walking—and the human family—a sense of humor is useful. So observed locomotion expert Henry McHenry of the University of California, Davis. He began a recent article on the origins of upright walking with this apocryphal story: When the late Tibetan Buddhist scholar and meditation master Chogyam Trungpa Rinpoche was asked by a student why our ancestors stood up on two legs, he replied: "A sense of humor."

A sense of humor may be as good an explanation as any for why our ancestors chose this curious form of locomotion, because behavior doesn't fossilize. "The first steps of our lineage will always be obscured from our full view," McHenry wrote. But indirect evidence about where the first footsteps were taken is coming from studies of the ancient environments where the earliest hominids lived and died. Indeed, the emerging lineup of early human ancestors has intensified research into the lost worlds they inhabited. These ambitious studies are providing an avalanche of data that is beginning to reveal the type of terrain where some chimp-sized primate took its first baby steps toward humanness. And the real Garden of Eden, it turns out, looked a lot more like the Okavango Delta than the wide-open spaces of the modern Great Rift Valley.

For one stifling week in August 2004, seven thousand earth scientists converged on Florence for the Thirty-second International Geological Congress. After the daytime scientific sessions ended, they fanned out over the ancient city's historic monuments like an army of red ants, identifiable by the red backpacks they had all been issued. They swarmed under Brunelleschi's famous dome for a Bach organ concert, completely filling the monumental Il Duomo cathedral for two nights in a row. But though impressive, the antiquities of the Renaissance were only so much recent history for many of these researchers who regularly delve into deep time, including those earth scientists whose mission is to reconstruct the ancient worlds where the earliest human ancestors and their relatives, the Miocene apes, lived millions of years ago.

Researchers have known for two decades that the earth's climate has been cooling over the past 50 million years, and that the cooling trend intensified during a cold, dry period 5.4 million to 6.5 million years ago. Less rain fell on Africa and Eurasia, and the Mediterranean Sea shrank—right at the time when the earliest human ancestors were presumably trying out and perfecting upright gaits. A key question is whether our ancestors, presumably well suited to their wooded habitats, were forced to make adaptations that sent them down the path toward humanness. If rain forests were dwindling, thickets of fruit trees might have been thinned into patches of forest, for example, and early human ancestors might have changed their locomotion to move from one patch of fruit trees to the next, with their hands free to carry food.

Some evidence has accumulated that woodland gave way to more grasslands between 6 million and 8 million years ago, and there was a major shift in the type of animals that lived south of the Sahara Desert in Africa 5 million to 7 million years ago. This is based on the ratios of isotopes of carbon in the teeth of large mammals around the world. But so far, no one has been able to convincingly link fluctuations in global climate and the thinning of forests on the African continent to changes in the local habitats where the fossils of the earliest human ancestors have been found in Chad and eastern Africa.

Ethiopian paleoanthropologist Yohannes Haile-Selassie reported in Florence that, despite attempts, he has found no evidence of a transition from a more wooded environment to open grasslands in the Middle Awash when *Ardipithecus kadabba* lived there 5.6 million to 5.8 million years ago. After a decade of collecting twenty-five hundred fossils of mammals and studying isotopes in the soils, he and geologist Giday WoldeGabriel have proposed that *Ardipithecus kadabba* lived in a wet forest alongside extinct pigs, elephants, and four different types of monkeys.

He did find evidence of a different type of transition: by comparing the species of animals at different fossil sites of similar age in Africa, Europe, and Asia, he found that some mammals were migrating between Europe, Asia, and Africa, presumably moving over a land bridge that connected what is now Yemen to Ethiopia 5.6 million years ago. Indeed, one of the questions currently being hotly debated is whether an ape that was the ancestor of the African apes and humans was ambling along with the animals on those migrations from Europe and Asia—or did the ancestor of African apes and humans arise in Africa (the prevailing view)?

Once in Africa, these immigrants fell in beside herds of native antelopes, elephants, equids, and other beasts moving along the rift valleys, which were giant corridors for migrating mammals. Indeed, in the Tugen Hills *Orrorin tugenensis* probably saw some of the same species of mammals when it was alive as did *Ardipithecus kadabba* in Kenya. *Orrorin* also lived in the woods, as did *Australopithecus anamensis,* who may have inhabited a gallery forest near the floodwaters at Kanapoi.

Those mammals on the move evidently also migrated between eastern and western Africa, because fossil sites in Chad and Libya share species with eastern African sites. At the meetings in Florence, a young French paleontologist who trained with Brunet, Jean-Renaud Boisserie, reported that the Mission Paléoanthropologique Franco-Tchadienne had collected ten thousand fossils of animals from four areas, including the Toros-Menalla area where Toumaï was found. The scenario emerging from studies of these bones showed that Toumaï dwelled in a gallery forest, living alongside at least one type of monkey whose jaw was found with it. Toumaï would have had to watch out for snakes, saber-toothed cats, and other predators. When it ventured out of the woods to the grassy shore of a nearby lake or marsh, it would have seen prehistoric crocodiles and hippos, turtles and otters, whose remains were found in the rich collection of fossils from the Toros-Menalla beds. Fossils of many fish that live in floodplains suggested that Toumaï lived near the shore or the shallow floodwaters of ancient Mega-Lake Chad, which once covered an area that was 400,000 square kilometers (it has shrunk to about 5,000 square kilometers today). The woods were a lush oasis, a fringe of green between the lake and a sandy desert—grains of sand found in fossil dunes showed that the sandy desert was nearby. It was the kind of landscape that could be found today in the Okavango Delta, Boisserie said.

~~~

On April 7, 2005, Toumaï was back on the cover of *Nature,* this time with an entirely new look. A new state-of-the-art reconstruction of the skull had inspired an artist to make a bust out of clay, to recall ancient creation myths that the first humans were made from clay. The skull wasn't pictured looking over the Djurab Desert, as it had been on its first cover of *Nature*. Instead, Toumaï was gazing over the swamplands of the Okavango Delta in Botswana.

Brunet and his colleagues had found fresh fossils—of a jaw and teeth of Toumaï's species, *Sahelanthropus*. But what put Toumaï back on the cover of *Nature* was a spectacular new reconstruction, using computational imaging to rebuild its distorted skull. On his first world tour with Toumaï, Brunet had stopped in Zurich to show the skull to a pair of researchers who are the acknowledged leaders in building three-dimensional, computer-aided reconstructions of fossils. The pair—neurobiologist Christophe Zollikofer and anthropologist Marcia Ponce de León of the University of Zurich—remove distortion from skulls that have been damaged, so that more accurate analysis can be done on those rebuilt skulls. When they took a look at the original skull of Toumaï (they do not work with casts), it was clear it was the best preserved skull of an ancient human ancestor ever found, with 95 percent of the skull's anatomy intact (minus its lower jaw). But its face was skewed after millions of years of being squashed under sediments. Their job was to erase the ravages of time and rebuild it to look like it had at the time of its death.

Working with Brunet, the pair began by making high-resolution CT scans of the skull, which took them five days. They ended up with images that broke up the skull into about one hundred different pieces. Working at separate computers so they would not see what the other one was doing, Zollikofer and Ponce de León then used three-dimensional computer graphics tools and software they had developed to rebuild the skull piece by piece, assembling the pieces into a patchwork where each piece fit precisely with all the pieces next to it and lined up along well-known fracture lines—a process constrained by geometry and anatomy. They did not compare their separate reconstructions until the end, to try to remove the possibility of bias.

As it turned out, they both ended up giving Toumaï a face-lift. Once they removed the distortion, the face was taller and its mug protruded a bit more than the original skull showed. They also identified thirty-nine landmarks on the skull, which are anatomical points of reference that they used to compare the skull directly with the skulls of other human ancestors, gorillas, and the two species of chimpanzees. They found that Toumaï's skull fell precisely within the shape of skulls of human ancestors. But no matter how hard they tried, they could not get it to fit the shape of a gorilla or chimpanzee skull without deforming it grossly.

Once they showed that the skull's shape was more like that of a human than that of a modern ape, Brunet had one more test for Toumaï. Was it possible to prove that a skull without a body had walked upright? For this, Brunet enlisted his former graduate student Franck Guy, who was a postdoctoral fellow at Harvard, and Harvard paleoanthropologist Dan Lieberman, who studies how the development of skulls and other bones is influenced by activity. Working with Zollikofer and Ponce de León, the team used landmark analysis to figure out the angle at which the skull had connected with the spine. They found that a plane at the base of Toumaï's neck formed a right angle to a virtual line drawn from the top to the bottom of its eye socket—an angle seen in humans that reflects the way our heads sit directly atop our vertical spines when we walk upright. The angle between the planes is much smaller in apes, which walk on all fours, reflecting that their heads jut out farther in front of their necks, which are more horizontal.

The team stopped short of declaring Toumaï an upright walker. It wrote in *Nature* that *Sahelanthropus* "might" have walked upright. But in interviews, the team members could think of no other reason for the restructuring of the base of the skull. "You need postcranial fossils to be one hundred percent sure, but it's darned hard to think how Toumaï could not have walked upright," said Lieberman.

Many paleoanthropologists agreed that Toumaï was not the ancestor of gorillas or chimpanzees—and the evidence was getting stronger that it was a hominid. But it was not an airtight case. Pickford regarded the new face of Toumaï overlooking the Okavango, and said, "I consider that he's wading across the same swamp that Louis Leakey floundered through with *Kenyapithecus,* and that Simons and Pilbeam got bogged down in with *Ramapithecus*—short face, small canine. Sound familiar?"

But even as paleoanthropologists debate the fine details of Toumaï's anatomy, Brunet has headed back to the source in the Djurab Desert for more evidence. He has found fossil beds with even older animals than those associated with Toumaï. But that is not enough: he also has staked out new terrain to the north, in Libya. He has a hunch that when ancient animals migrated between Libya and Chad, some little two-legged ancestors might have been among the camp followers that moved along wooded corridors linking the two regions. The nomadic French paleontologist confided this idea to Libyan president Muammar Qaddafi on an official French expedition to Libya recently. Brunet left Qaddafi's tent with the name of a guide and permission to explore southern Libya. Soon after, he made a new bet with David Pilbeam at Harvard—that he would find early human ancestors in Libya. Pilbeam thinks Libya is too far north to have supported forest-dwelling hominids, but he has learned to trust Brunet's instincts. "Life is full of surprises," Pilbeam said. Brunet, who has a formidable track record of winning bets and fulfilling his promises, had a response for skeptics who said he needed skeletal bones to prove that Toumaï was the first member of the human family. He said cryptically: "More will be coming."

ACKNOWLEDGMENTS

I would like to express my deep gratitude to all of the people in the paleoanthropological profession who have talked with me, opened their camps to me, given me tutorials, and responded to my many requests for interviews, manuscripts, copies of letters, documents, and photos over the past three years. I could not have written this book without the generous cooperation of many researchers, even though at times my inquiries were inconvenient or distressing.

This book is not a comprehensive history. It is my perception of the quest for the earliest ancestors during the past fifteen years, as I covered the science of human origins for *Science*. I have focused on the leaders of four teams that found the earliest known members of the human family. There are many other researchers whose work on the origins of humans is equally worthy but whom I could not mention for the sake of the narrative. At the outset, I was reminded of the contentiousness of the field, and some researchers were wary when I told them I would be talking with their rivals. Indeed, some researchers urged me to stick to the science and to avoid writing about the politics and personal rivalries, because it might reflect poorly on the field and provide ammunition for creationists trying to malign Darwinian evolution and the scientists who have dedicated their careers to understanding it. I found it impossible, however, to separate the human story of the quest from the scientific results; science is a social endeavor and the personal politics influence not only who gets access to data, in the form of fossils and fossil sites, but even how researchers interpret the fossils and formulate hypotheses. In the end, I decided to include personal details where they influenced the science or revealed the motivations of the scientists. My intent was to show the triumphs of the science of paleoanthropology and Darwinian evolution in the past century, despite personal battles and intense rivalries, false starts and mistakes. The science lurches forward despite the foibles of the individual scientists.

Although I am indebted to many researchers, I owe special thanks to several teams who gave me unrestricted access to their camps or teams. In particular, I thank Meave Leakey and Fredrick Kyalo Manthi, who invited me to travel to Kanapoi with them, as well as Justus Edung, Benson Maina Gachaga, and Robert Moru of the National Museums of Kenya, who tolerated my efforts to find and sift fossils. I also thank Martin Pickford and Brigitte Senut for hosting me in their camp in the Tugen Hills, and for showing me how they search for fossils and answering many inquiries. I am deeply grateful to Michel Brunet for allowing me to follow him for several days in Poitiers, initially for a profile in *Science*, as well as for granting me interviews and responding to countless requests for more information and photographs. I also thank Berhane Asfaw, Yohannes Haile-Selassie, Bruce Latimer, Scott Simpson, Tim White, and Giday WoldeGabriel for interviews in Berkeley, Cleveland, Florence, and Paris and for the use of years of notes from past interviews for *Science*. Andrew Hill tolerated repeated interviews, as well.

I owe a tremendous debt to many researchers who read sections of the manuscript. In particular, I thank David Pilbeam and Tim White for their thorough and thoughtful comments on large sections of an early, partial draft of the manuscript. I am, of course, entirely responsible for any errors and interpretations.

I am extremely grateful for the cooperation and information from the following people: Stanley Ambrose, Peter Andrews, Alain Beauvilain, Anna K. Behrensmeyer, Raymond Bernor, Katie Binetti, René Bobe,

Jean-Renaud Boisserie, Raymond Bonnefille, Alison Brooks, Frank Brown, Yves Coppens, Gerald Eck, Robert Eckhardt, Idle Farah, Craig Feibel, Michael Fortelius, Henry Gee, Eustace Gitonga, Fanoné Gongdibé, Morris Goodman, Terry Harrison, F. Clark Howell, Donald Johanson, Jon Kalb, Jay Kelley, Mzalendo Kibunja, Bill Kimbel, Richard Klein, Jane Kyaka, Richard Leakey, Dan Lieberman, Owen Lovejoy, Laura MacLatchy, Emma Mbua, James Ohman, Marcia Ponce de León, Rick Potts, Paul Renne, Lorenzo Rook, James B. Rossie, Vince Sarich, Robert Saunders, Jeff Schwartz, Horst Seidler, Sileshi Semaw, William Sill, Elwyn Simons, Maurice Taieb, Ian Tattersall, Alan Walker, Carol Ward, Gerhard Weber, Milford Wolpoff, Craig B. Wood, Roger Wood, Richard Wrangham, John Yellen, Adrienne Zihlman, and Christophe Zollikofer.

The idea for this book took shape when *Science* assigned me to profile Michel Brunet. I owe a tremendous debt to my editor, Elizabeth Culotta, for sending me on that assignment and for being a terrific editor and friend, as well as for her deft read of sections of my manuscript. Colin Norman at *Science* also has been the best kind of editor for many years. I am doubly indebted to my longtime colleague Michael Balter, for encouraging me to write this book, and for his constant support and friendship, as well as for introducing me to my agent, Heather Schroder, whose enthusiasm and advice have been critical for the success of this project. At Doubleday, I am deeply grateful to my editor, Charles Conrad, for his guidance and thoughtful editing, to Alison Presley for keeping me organized in an encouraging way, and to Bonnie Thompson for her thorough copyedit. I thank Karen Meyers and David Brill for their high-quality photographs, and Doug Millar and his associates Otto Beck and Dave Harr for their travel arrangements and contacts in Kenya.

Finally, I thank Nola Gibbons, Barbara Scherlis, Leonard Scherlis, Julie Leftwich, Judy Christenson, and Susan Sternburg for their patience, friendship, and constant support. Special thanks to John Scherlis for many thought-provoking discussions. This project could not have been completed without the assistance and friendship of Tracy Kelly Seighman, Brandi Giering, and Mary Ann Mervis. To my children, Lily, Sophia, and Tom, thank you for so many sacrifices and for tolerating Mom's working on nights and weekends and traveling to distant places. For my husband, Bill Scherlis, thank you for just about everything—from your red ink and graphic designs to your technical and emotional support.

GLOSSARY

ADAPTATION: the process by which an organism becomes adjusted to its environment, such as responding to long-term environmental stresses through permanent genetic change, i.e., natural selection or evolving.

ADAPTIVE RADIATION: the relatively rapid expansion and diversification of an evolving group of organisms as they adjust to new environments or ways of life.

AFRICAN APES: chimpanzees and gorillas.

ANTHROPOID: a member of the primate suborder Anthropoidea. Monkeys, apes, and humans are anthropoids.

ARCHAEOLOGY: the study of the material remains of past cultures and peoples, such as stone tools and artifacts.

ARGON-ARGON DATING: a radiometric dating method based on the changing ratio of argon-40 to argon-39 with the passage of time in volcanic rock or ash. This technique was derived from potassium-argon dating. The argon-argon method is usually more accurate than potassium-argon dating and doesn't require as large a sample.

AUSTRALOPITHECINES: a subfamily (Australopithecinae) consisting of the single genus *(Australopithecus)* of extinct hominids that lived from about 4.1 million years ago until about 1 million years ago. Australopithecines were characterized by upright walking, robust teeth and jaws, and ape-sized brains. Most taxonomists recognize six species: *A. anamensis, A. afarensis, A. africanus, A. aethiopicus, A. robustus,* and *A. boisei.*

BIPEDALISM: walking on the hind limbs, especially in an upright, human manner.

CONVERGENCE: the parallel development of the same feature in unrelated organisms either by chance or because of independent adaptation to similar environments or ways of life, such as wings in birds and bats. Also called convergent evolution. It contrasts with homologous evolution.

DARWINIAN THEORY OF EVOLUTION: the theory proposed by Charles Darwin in *The Origin of Species* (1859). Animals and plants in a population inherit variations in traits from their ancestors; more individuals are born than can survive (a struggle for existence); and individuals carrying certain favorable variants of traits are more likely to survive and pass them on to others (natural selection). Over long periods of time, natural selection gives rise to new forms of life (the origin of species).

DARWINISM, NEO-DARWINISM: evolution through natural selection. Neo-Darwinism is a fusion of the Darwinian theory of evolution with the genetic mechanisms of heredity; in other words, variation arises by mutations at the genetic level that, if advantageous, can be retained and reinforced (through natural selection) in future generations.

DERIVED CHARACTER: a new trait developed in a more recent ancestor and retained by descendants but absent in older ancestral stock, which shows a primitive version of the same trait.

DNA: deoxyribonucleic acid, the molecule that carries hereditary information in all living cells. DNA sequence is the order of bases along the DNA chain that carries the information needed to code for synthesizing proteins.

FOSSIL: preserved traces or remains of past life more than 10,000 years old, embedded in rock as mineralized remains or as impressions, casts, or tracks.

GENUS (plural GENERA): a taxonomic category above the species level but below the family. There may be several species in a genus, such as the genus *Homo*, which includes *Homo erectus* and *Homo sapiens*, among others. There can be several genera in a family, such as the genera of *Homo*, *Australopithecus*, and *Ardipithecus* in the human family, Hominidae.

GREAT APES: chimpanzees, gorillas, and orangutans.

HOLOCENE EPOCH: the epoch from 10,000 years ago up through the present.

HOMINID: a creature belonging to the family of primates (Hominidae, also called Homininae, depending on the method of classification) of which *Homo sapiens* is the only surviving species. Also includes ancestors of humans, including all species of *Homo* and *Australopithecus*. In this book, it is the term used to describe humans and their ancestors, and it does not include chimpanzees, gorillas, or other apes.

HOMININ: many researchers have recently begun to classify humans and their extinct ancestors in the family Homininae, as opposed to the Hominidae (or hominids). This is because genetic studies done in the 1990s showed that chimpanzees are so closely related to humans that many researchers now include chimpanzees in the family Hominidae, with humans. Therefore, humans are designated by their own separate tribe or subfamily, Homininae—in common terms, *hominin*, rather than *hominid*.

HOMINOID: of or belonging to a superfamily (Hominoidea) of primates that includes apes and humans.

HOMO: the genus that includes humans (named in 1758 by Linnaeus) and some of their ancestors. *Homo sapiens* is the only surviving species. Members of the genus *Homo* are characterized by large brains, straight (rather than projecting) facial profiles, and other traits in the teeth and skull. Most taxonomists recognize at least five species: *H. habilis, H. erectus,* archaic *H. sapiens, H. neanderthalensis,* and *H. sapiens;* other species have been proposed.

HOMOLOGYOUS EVOLUTION: the development of traits that are similar because of common ancestry, not convergence.

HYLOBATIDAE: the family of lesser apes that includes gibbon apes and their ancestors.

ISOTOPE: any of two or more forms of an element that have the same chemical properties and the same atomic number but different atomic weights and slightly different physical properties because their nuclei have different numbers of neutrons. Most elements occur naturally as mixtures of isotopes.

KNUCKLE-WALKING: a type of four-legged locomotion used by chimpanzees and gorillas where they support the weight of their bodies on the back of their knuckles.

MIOCENE EPOCH: the epoch from 5.3 million to 23.8 million years ago. Includes the period when the first apes arose and when the common ancestor of humans and chimpanzees split.

MOLECULAR CLOCK: the hypothesis based on the assumption that mutations accumulate at a steady rate over long periods of time and thus can be used to estimate the time elapsed since two species split from a common ancestor.

MONKEYS: long-tailed primates, excluding apes, humans, and prosimians.

NATURAL SELECTION: the differential survival and reproduction of individuals with different genotypes, or versions of genes, within a population.

PALEOANTHROPOLOGY: the study of the human fossil record and archaeology.

PALEONTOLOGY: the study of the fossilized remains of plants and animals.

PHYLOGENY: the evolutionary relationships among a group of species, usually diagrammed as a family tree, or phylogenetic tree. The relationships can be determined by studies of fossils or genetics.

PLEISTOCENE EPOCH: the epoch before the present period, from just under 2 million years to 10,000 years ago. During this time the first members of the genus *Homo* appeared in Africa.

PLIOCENE EPOCH: the latest epoch of the Tertiary period, from about 5.3 million to just under 2 million years ago.

PONGIDAE: the family of great apes that includes chimpanzees, gorillas, and orangutans and their ancestors.

POSTCRANIAL BONES: bones of the skeleton excluding the skull.

PRIMITIVE CHARACTER: a trait inherited from the common ancestor of a group of species; also known as an ancestral trait. Primitive characters are contrasted with derived characters.

PROTEINS: large molecules that consist of a long chain of amino acids. Proteins perform special functions, such as in immune response or physical development.

QUADRUPEDALISM: walking on four legs.

QUARTERNARY PERIOD: the period of the Cenozoic era that includes the Holocene and Pleistocene epochs (from just less than 2 million years ago to the present).

RADIOMETRIC DATING: an absolute dating method based on the known rate of decay of isotopes, such as potassium to argon and carbon to nitrogen.

SPECIES: a group of organisms classified together at the lowest level of the taxonomic hierarchy. Biologists disagree about how to define a species, but many accept the biological species concept, which says that two groups of organisms belong to different species if they cannot interbred in nature and produce fertile offspring.

TECTONICS: the study of the earth's crustal structures, such as continental plates, and the forces that cause them to change shape and move relative to one another.

TERTIARY PERIOD: the first period of the Cenozoic era. It includes the Pliocene, Miocene, Oligocene, Eocene, and Paleocene epochs (65 million to just under 2 million years ago).

NOTES

Epigraph

vii   "It has been said, that the love of the chase": Charles Darwin, *Diary of the Voyage of H.M.S. Beagle*, in vol. 1 of *The Works of Charles Darwin* (New York: New York University Press, 1987).

Introduction

1   The main sources for Michel Brunet's background, quotes, and discovery of the fossil jaw-bone Abel are the author's interviews with Brunet in Poitiers, France, October 1–3, 2002, and telephone interviews on November 14, 2001, and April 2, 2005. Other sources include: Brunet, "Dreams of the Past," *Nature* 423 (2003), p. 121, and Jean-Luc Terradillos, "Abel: L'Homme de la Rivière aux Gazelles," *L'Actualité Poitou-Charentes* 31 (1996), pp. 16–21.

2   "Those little rodents live out in the dry places": John Mangels, "Fossil Hunter Transcripts," Cleveland *Plain Dealer*, November 28, 2004; e-mail from Tim White on June 26, 2005.

2   "I am working in older sediments": Michel Brunet, interview in Poitiers, October 3, 2002.

4   Mamelbaye Tomalta, called Brunet to come over: Alain Beauvilain, telephone interview, December 2, 2004.

4   He said: "David, I've got it": David Pilbeam, telephone interview, October 2002.

10   The main sources for Tim White's background and quotes are the author's telephone interviews and e-mails with White on June 30, 2003; July 2, 2003; November 6, 2004; and June 26, 2005. Other sources include: Tim White, "At Large in the Mountains," in *Curious Minds: How a Child Becomes a Scientist*, ed. John Brockman (New York: Pantheon Books, 2004); and an interview with Tim White by Harry Kreisler, "On the Trail of Our Human Ancestors," September 18, 2003, Conversations with History, Institute of International Studies, University of California at Berkeley; http://globetrotter.berkeley.edu/people3/White/white-con1.html.

11   "Tim Commandments": Tim D. White, "A View on the Science: Physical Anthropology at the Millennium," *American Journal of Physical Anthropology* 112 (2000); pp. 287–92.

11   When asked on National Public Radio: Interview with Tim White on National Public Radio, *Talk of the Nation/Science Friday,* March 19, 2004.

14   Information about the discovery of the skeleton of *Ardipithecus ramidus* was based on the author's interviews: with Berhane Asfaw, by telephone August 6, 2002; Yohannes Haile-Selassie, in Cleveland, Ohio, March 10, 2004; Tim White, by e-mail, on November 10, 2004; and Scott Simpson, in Cleveland, on November 22, 2004.

15   The main sources for Meave Leakey's background are the author's interviews with Leakey on October 20–31, 2003, in Kenya and by e-mail on July 2, 2005. Other sources include: Virginia Morell, *Ancestral Passions: The Leakey Family and the Quest for Humankind's Beginnings* (New York: Simon & Schuster, 1995).

15   The main sources for Meave Leakey's meeting with Tim White in Addis Ababa are the author's interviews with Leakey on October 19, 2003, en route to Kanapoi, Kenya, and e-mail on July 2, 2005; and e-mail from White on June 26, 2005.

18   His decisions—and autocratic approach: David Western, "Wildlife Conservation in Kenya," *Science* 280 (1998), p. 1507 (in "Letters").

20   Meave and paleontologist Alan Walker: Meave Leakey, Craig S. Feibel, Ian McDougall, and Alan Walker, "New Four-Million-Year-Old Hominid Species from Kanapoi and Allia Bay, Kenya," *Nature* 375 (1995), pp. 565–71.

21   "We're on tenterhooks": John Noble Wilford, "In Kenya, Fossils Are Found of Earliest Walking Species," *New York Times,* August 17, 1995, p. A1.

Part One: Ancient Footsteps

23   "I hear the ancient footsteps": Bob Dylan, "Every Grain of Sand," *Shot of Love* (Los Angeles: Columbia Records, 1981).

Chapter One: African Trailblazers

25   "Most scientific problems are far better understood": Ernst Mayr, *The Growth of Biological Thought* (Cambridge: Belknap Press–Harvard University Press, 1982), p. 6.

25   "I was told as a young student": L. S. B. Leakey, *White African* (New York: Ballantine Books, 1966), p. x.

26   But Louis's search for the missing link: The main sources for information about Louis Leakey's life and work are: interview by the author with Meave Leakey at Kariandusi on October 20, 2003; L. S. B. Leakey, *White African;* L. S. B. Leakey, *By the Evidence* (New York: Harcourt Brace Jovanovich, 1974); Virginia Morell, *Ancestral Passions.*

28   Charles Darwin had proposed: Charles Darwin, *The Descent of Man and Selection in Relation to Sex* (London: John Murray, 1871).

28   Ernst Haeckel, believed that the Asian apes: Ernst Haeckel, *The History of Creation, or The Development of the Earth and Its Inhabitants by the Action of Natural Causes: A Popular Exposition of the Doctrine of Evolution in General, and That of Darwin, Goethe, and Lamarck in Particular,* trans. E. Ray Lankester (New York: D. Appleton, 1868).

29   Eugène Dubois, who became the first: The main sources for Eugène Dubois's life and work are: Pat Shipman, *The Man Who Found the Missing Link* (New York: Simon & Schuster, 2001), and Alan Walker and Pat Shipman, *The Wisdom of the Bones* (New York: Random House, 1996).

31   "We must, however, acknowledge": Charles Darwin, *The Descent of Man.*

31–32   the Piltdown hoax: Stephen Jay Gould, "Piltdown Revisited," in *The Panda's Thumb* (New York: W. W. Norton, 1980).

33   When he got to Johannesburg: The main sources for Raymond Dart's discovery of the Taung baby are: Raymond A. Dart with Dennis Craig, *Adventures with the Missing Link* (New York: Harper & Brothers, 1959); Walker and Shipman, *Wisdom of the Bones.*

37   Louis later wrote that: L. S. B. Leakey, *White African,* pp. 68–73.

39   "The discovery was, therefore": L. S. B. Leakey, *White African,* p. 201.

41   Mary Leakey discovered Zinj: Mary Leakey, *Disclosing the Past: An Autobiography* (New York: Doubleday, 1984), pp. 120–21.

42   The Leakey biographer Virginia Morell: Virginia Morell, *Ancestral Passions.*

43   (that date of 1 million years had doubled since the 1920s): William Glen, *The Road to Jaramillo: Critical Years of the Revolution in Earth Science* (Stanford: Stanford University Press, 1982).

43   Jack Evernden and Garniss Curtis had traveled to Olduvai: William Glen, *The Road to Jaramillo,* pp. 75–77.

44   "It was four times older": William Glen, *The Road to Jaramillo,* p. 77.

45   "*Zinjanthropus* had come into our lives": Mary Leakey, *Disclosing the Past,* pp. 121, 140–42.

Chapter Two: Continental Divide

46   "The exceedingly cut-throat level": Robert Bell, *Impure Science: Fraud, Compromise, and Political Influence in Scientific Research* (New York: John Wiley & Sons, 1992), p. 31.

46   The main sources for Bryan Patterson's life and discoveries in Kenya are the author's interviews with his former students, including Anna K. Behrensmeyer on January 14, 2004, and in Washington, D.C., on January 19, 2004; and telephone interviews with William Sill on January 27, 2004; Roger C. Wood on January 27, 2004; and Craig B. Wood on January 28, 2004. William Sill also provided copies of his movies and field notes from Patterson's expeditions in Kenya in 1964, 1965, and 1967. Also: "Professor Bryan Patterson Retires," *Museum of Comparative Zoology Newsletter* 4 (Spring 1975); William D. Turnbull and Farish A. Jenkins Jr., "Bryan Patterson, 1901–1979," *Society of Vertebrate Paleontology News Bulletin* (February 1981).

47   "Ho hum, there's another knucklebone": Walter Sullivan, "Bone Found in Kenya Indicates Man Is 2.5 Million Years Old," *New York Times,* January 14, 1967.

48   "none of us was willing to undergo": Roger C. Wood, in a telephone interview, January 27, 2004.

49   "But you were not supposed to find hominids!": William Sill, in a telephone interview, January 27, 2004; entry in William Sill's field notes, August 25, 1965.

49   he was not surprised because he never had found: Roger C. Wood, in a telephone interview, January 27, 2004.

49   A new team came to eastern Africa each year: Yves Coppens, "East Side Story: The Origin of Humankind," *Scientific American,* May 1994, pp. 88–95.

53   The main sources for the Omo Expedition were the author's telephone interviews with Clark Howell on October 26, 2004; Frank Brown on January 9, 2004; and Yves Coppens on April 18, 2005, and, in Paris, on October 5, 2002.

54   On June 3, 1967, a convoy of three trucks: Virginia Morell, *Ancestral Passions,* p. 282.

55   Howell was grumpy about it: Frank Brown, telephone interview on January 9, 2004.

58   "worst offenders are Geologists": Virginia Morell, *Ancestral Passions;* Roger C. Wood, in a telephone interview, January 27, 2004; Anna K. Behrensmeyer, telephone interview, January 14, 2004.

Chapter Three: The Early Ancestor

59   "Where, then, must we look": Thomas Henry Huxley, *Evidence as to Man's Place in Nature* (New York: D. Appleton, 1863).

59   Patterson's fossil elbow: Walter Sullivan, "Bone Found in Kenya."

59   "Man's separation from his closest cousins": Lawrence Fellows, "Man's Ancestor Found to Be More Than 19 Million Years Old," *New York Times*, January 15, 1967.

60   the earliest members of the human family: L. S. B. Leakey, *By the Evidence*, p. 230.

60   many of his scientific colleagues were skeptical: Elwyn Simons, in e-mail, July 15, 2005.

61   The main sources for the diversity of extinct apes in the Miocene were Jay Kelley and Andrew Hill, in interviews in Chicago on October 11, 2003. Other sources include Steve Ward and Dana Duren, "Middle and Late Miocene African Hominoids," and David Pilbeam, "Perspectives on the Miocene Hominoidea," both in *The Primate Fossil Record*, ed. Walter Hartwig (New York: Cambridge University Press, 2002).

64   The main sources for the discovery of *Ramapithecus* were Elwyn Simons, in telephone interviews on September 7, 2004, and October 18, 2004, and David Pilbeam, interview at Harvard University, July 22, 2004. Two books discuss this episode in detail: Roger Lewin, *Bones of Contention* (New York: Simon & Schuster, 1987), and Ian Tattersall, *The Fossil Trail* (New York: Oxford University Press, 1995).

67   an approach known as the New Systematics: Mayr, *Growth of Biological Thought*, pp. 276–79.

67   "biological species concept": Mayr, *Growth of Biological Thought*, pp. 273, 279–86.

68   now dated from 8 million to 13 million years ago: David Pilbeam, telephone interview, June 8, 2005.

68   "Hands were probably used extensively": Elwyn Simons and David Pilbeam, "Preliminary Revision of Dryopithecinae (Pongidae, Anthropoidea)," *Folia Primatologica* (1965), pp. 81–152; David Pilbeam, "Notes on *Ramapithecus*, the Earliest Known Hominid, and *Dryopithecus*," *American Journal of Physical Anthropology* 25 (1966), p. 2.

69   It was shown walking upright: F. Clark Howell and the Editors of Life, *Early Man* (New York: Time, 1965), p. 42.

69   "We can safely put this specimen": Bernard Campbell, *Human Evolution* (Chicago: Aldine Publishing Co., 1966), p. 94.

Chapter Four: Drawing Bloodlines

70   "Discoveries made in a field": Craig J. Venter, "Edge: The World Question Center," *Edge.org*, 1994, www.edge.org/q2004/p.5.html.

70   "What we know about human origins": Alan Walker, in a talk at the Leakey Foundation Centennial Tribute to Louis Leakey, Chicago, October 11, 2003.

70   Vincent Sarich was a graduate student: Vince Sarich, telephone interview, October 28, 2004; interview in Tampa, Florida, on April 15, 2004.

71   "Of course I like to raise hell": Paul Selvin, "The Raging Bull of Berkeley," *Science* 251 (1991), p. 368.

71   thought that humans were sufficiently similar: S. L. Washburn, ed., *Classification and Human Evolution* (Chicago: Aldine, 1963).

71   Washburn had been influenced: Jonathan Marks, "Sherwood Washburn, 1911–2000," *Evolutionary Anthropology* (2000), p. 225.

72   Goodman and a colleague: Morris Goodman, telephone interview, January 25, 2005.

73   "Naively, I anticipated": Morris Goodman, "Epilogue: A Personal Account of the Origins of a New Paradigm," *Molecular Phylogenetics and Evolution* 5 (1996), pp. 269–85.

73   "Washburn and I were alike": Vince Sarich, telephone interview, October 28, 2004.

74   Zuckerkandl and Pauling had been the first to show: Sarich, telephone interview, October 28, 2004, and interview in Tampa, Florida, on April 25, 2004. Also in Roger Lewin, *Patterns in Evolution: The New Molecular View* (New York: Scientific American Library, 1999), and Vincent Sarich, "Immunological Evidence on Primates," in *The Cambridge Encyclopedia of Human Evolution*, eds. Steve Jones, Robert Martin, and David Pilbeam (Cambridge: Cambridge University Press, 1992), pp. 303–06.

74   an overwhelming number of studies have since shown: Ann Gibbons, "What Genes Make Us Human?" *Science* 281 (1998), pp. 1432–34; Ann Gibbons, "When It Comes to Evolution, Humans Are in the Slow Class," *Science* 267 (1995), pp. 1907–08; Michael Steiper, Nathan M. Young, and Tika Y. Sukarna, "Genomic Data Support the Hominoid-Cercopithecoid Divergence," *Proceedings of the National Academy of Sciences*, November 9, 2004; The Chimpanzee Sequencing and Analysis Consortium, *Nature* 437 (2005), pp. 69–87.

74   By the end of 1967, Sarich and Wilson: Vincent Sarich and A. C. Wilson, "Immunological Time Scale for Hominid Evolution," *Science* 158 (1967), pp. 1200–03.

75   "man and African apes shared a common ancestor": Vincent Sarich and A. C. Wilson, "Immunological Time Scale."

75   "not in accord with the facts available today": Louis Leakey, "The Relationship of African Apes, Man, and Old World Monkeys," *Proceedings of the National Academy of Sciences* 67 (1970), pp. 746–48. Also Roger Lewin, *Bones of Contention*.

75   "It is not presently acceptable that *Australopithecus*": Elwyn Simons, "The Origin and Radiation of the Primates," *Annals of the New York Academy of Sciences* 167 (1968), pp. 319–331. Also Roger Lewin, *Bones of Contention*.

75   The molecular dates had not been proven: Elwyn Simons, telephone interview, September 7, 2004.

75   "One no longer has the option": Vince Sarich, and P. J. Dolhinow, eds. "A Molecular Approach to the Question of Human Origins," in *Background for Man* (Boston: Little Brown, 1971), pp. 60–81. Also Roger Lewin, *Bones of Contention*.

76   Pilbeam led an international expedition: David Pilbeam, interview in Cambridge, Massachusetts, July 22, 2004.

76   Pilbeam . . . the same fossil beds: David Pilbeam, "The Descent of Hominoids and Hominids," *Scientific American*, March 1984, pp. 84–96.

77   "Why was the hominoid fossil record misinterpreted": David Pilbeam, "The Descent of Hominoids."

77   thick tooth enamel . . . a small canine: David Pilbeam, "Distinguished Lecture: Hominoid Evolution and Hominoid Origins," *American Anthropologist* 88 (1986), pp. 295–312.

78   dates that still hold for most molecular anthropologists: Ann Gibbons, "When It Comes to Evolution."

78   "searchers were looking in the wrong place": Elwyn Simons, "Human Origins," *Science* 245 (1987), pp. 1343–50.

Chapter Five: Lucy, the Late Ancestor

79   "Time that was so long grows short": Anne Sexton, "For Mr. Death Who Stands with His Door Open," *The Death Notebooks* (Boston: Houghton Mifflin, 1974).

79   He was not the first geologist: Jon Kalb, *Adventures in the Bone Trade* (New York: Copernicus Books, 2001), pp. 32, 12.

80   By December 1969: Maurice Taieb, telephone interview, November 23, 2004.

81   "an enormous, flat-lying encyclopedia": Jon Kalb, *Adventures*, p. 84.

82   Coppens took a look at the elephant tooth: Yves Coppens, telephone interview, April 18, 2005.

83   Louis died the following October: Virginia Morell, *Ancestral Passions*, p. 402.

83   "It really took my breath away": Don Johanson, telephone interview, January 26, 2005.

84   Johanson was preoccupied with these worries: Donald Johanson and Maitland Edey, *Lucy: The Beginnings of Humankind* (New York: Simon & Schuster, 1981), p. 155.

85   He had studied the legs: Owen Lovejoy, "Proximal Femoral Anatomy of *Australopithecus*," *Nature* 235 (1972), pp. 175–76.

85   Lovejoy remembers the day: Owen Lovejoy, interview in Kent, Ohio, November 22, 2004.

86   the team would find about 20 to 40 percent: Alan Walker and Pat Shipman, *The Wisdom of the Bones*, p. 181.

Chapter Six: Defining Humans

88   "It may seem ridiculous for science": Don Johanson and Maitland Edey, *Lucy*, p. 100.

90   "smooth, young hotshot": Donald Johanson and Maitland Edey, *Lucy*, p. 218.

90   Johanson had been born in Chicago: Don Johanson and Maitland Edey, *Lucy*, p. 72.

91   "luckiest kid in the world": Tim White, "At Large in the Mountains," *Curious Minds*, p. 209.

92   White had badgered Wolpoff to ask the Leakeys: Tim White, telephone interview, July 2, 2003.

92   "When I stepped off the plane at Koobi Fora": Tim White, telephone interview, July 2, 2003.

92   he was also the beneficiary of Richard and Mary Leakey's: interest: Virginia Morell, *Ancestral Passions*, pp. 76–77.

95   This marked the beginning of the end: Virginia Morell, *Ancestral Passions*, p. 481.

95   Documentary filmmakers recorded the session: *Lucy in Disguise*, television documentary produced by WVIZ and the Cleveland Museum of Natural History (1980).

96   A few notable skeptics, including her codiscoverer Yves Coppens: Yves Coppens, *Lucy's Knee* (Pretoria: Protea Book House, 2002), pp. 134–37.

97   "I think that Lucy was the most important discovery": Vince Sarich, telephone interview, October 28, 2004.

Chapter Seven: Banishment

98   "Eating the bitter bread of banishment": William Shakespeare, *Richard II*, act 3, scene 1, line 21.

99   he tells the dramatic story of checking out fossils: Don Johanson and Maitland Edey, *Lucy*, p. 233.

100   "What we find in them could well blow": Don Johanson and Maitland Edey, *Lucy*, p. 375.

101   Kalb's time in the Middle Awash came to an abrupt end: John Kalb, *Adventures*, p. 280; Jon Kalb, telephone interview, February 24, 2005.

101   He adamantly denies that he worked for the CIA: Jon Kalb, telephone interview, February 24, 2005.

101   Kalb won an out-of-court settlement: Eliot Marshal, "Gossip and Peer Review at the NSF," *Science* 238 (1987), p. 1502; John Yellen, telephone interview, September 17, 1990.

101   program officers had gossiped about Kalb's rumored connection: Robert Bell, *Impure Science*, pp. 21, 29; Eliot Marshall, "Gossip and Peer Review"; and Jon Kalb, *Adventures*, p. 146.

101   Kalb later wrote that some of the researchers: Jon Kalb, telephone interview, February 24, 2005; also Jon Kalb, *Adventures*, p. 146.

101   "The rumor has no basis in fact": Robert Bell, *Impure Science*, p. 14.

101   "The exceedingly cut-throat level of competition": Robert Bell, *Impure Science*, p. 31.

102   A senior collaborator of Kalb's had already invited Clark: Jon Kalb, *Adventures*, p. 254; Tim White, e-mail, June 26, 2005; Roger Lewin, "Ethiopia Halts Prehistory Research," *Science* 219 (1983), pp. 147–49.

102   Johanson was charged with trying to steal a fossil: Maurice Taieb, telephone interview, February 27, 2005; Donald Johanson, e-mail, July 1, 2005.

102   Taieb and Johanson say the charge was ridiculous: Don Johanson, e-mail, July 1, 2005.

102   "I got tired of the politics": Maurice Taieb, telephone interview, February 27, 2005.

103   By that time, the fossils of hominids and artifacts: Tim White, telephone interview, July 2, 2003.

103   Kalb's former students who were Ethiopians: Roger Lewin, "Ethiopia Halts"; also Jon Kalb, telephone interview, February 24, 2005.

104   "I always said it was a hominid": Martin Pickford, interview in Nairobi, Kenya, September 15, 2003.

104–05   the report written with the paleoanthropologist Peter Andrews: Peter Andrews, e-mail, February 24, 2005.

105   trying to resist jumping into the "rat race": Martin Pickford, interview in Nairobi, Kenya, September 15, 2003.

106   The two were friends, as were their brothers: Martin Pickford, interview in Nairobi, Kenya, September 15, 2003; Richard Leakey, e-mail, June 17, 2005.

106   "We were alike. We were into animals": Martin Pickford, interview in Nairobi, Kenya, September 15, 2003.

106   It also taught him that he could not count on other people: Martin Pickford, interview in Nairobi, Kenya, September 15, 2003.

106   Pickford blames their rough start: Martin Pickford, interview in Nairobi, Kenya, September 15, 2003; telephone interview, October 4, 2005.

106   "He wasn't the kind of person I liked": Andrew Hill, telephone interview, April 3, 2005.

108   "a habit" of Pickford's: Jan van der Made, "Methods in Biostratigraphy: A Reply to Pickford et al.," *Geobios* 36 (2003), pp. 223–28.

108   "I am not the devil": Martin Pickford, interview in the Tugen Hills, Kenya, September 19, 2003.

109   "boring talk on mud in the middle of England": Andrew Hill, interview in New Haven, Connecticut, February 4, 2004.

110   "We need to reconstruct past species": David Pilbeam, "Distinguished Lecture."

111   But Ogot was forced to resign in 1980: Bethwell A. Ogot, *My Footprints on the Sands of Time* (Canada: Trafford Publishers, 2003), pp. 344–66; Martin Pickford, telephone interview, October 4, 2005.

111   recommended that the attorney general seek: David Pilbeam, e-mail, October 6, 2005.

111   the Lukeino molar and a fragment of an arm bone: M. Pickford, D. C. Johanson, C. O. Lovejoy, T. D. White, and J. L. Aronson, "A Hominoid Humeral Fragment from the Pliocene of Kenya," *American Journal of Physical Anthropology* 60 (1983), pp. 337–46.

111   Pickford had had "difficulties": David Pilbeam, e-mail, October 6, 2005.

111   would become a sore point for Pickford: Eustace Gitonga and Martin Pickford, *Richard E. Leakey: Master of Deceit* (Nairobi: White Elephant Publishers, 1995), p. 56; Martin Pickford, telephone interview, October 4, 2005.

111   Pilbeam denies this: David Pilbeam, telephone interview, July 22, 2005.

112   Hill wrote that the new fossil from Tabarin: Andrew Hill, "Early Hominid from Baringo, Kenya," *Nature* 315 (1985), pp. 222–24.

112   his proposal to focus there further was shelved: Andrew Hill, interview in New Haven, Connecticut, February 4, 2004.

113   But he was told by Leakey not to collect the fossils: Eustace Gitonga and Martin Pickford, *Master of Deceit*, p. 18.

113   did not treat her with sufficient respect after a guest lecture: Brigitte Senut, interview in the Tugen Hills, Kenya, September 16, 2003.

113   (White recalled the incident differently: Tim White, e-mail, June 26, 2005.

113   her mentor, Coppens, compared her to Joan of Arc: Yves Coppens, telephone interview, April 18, 2005.

114   Hill, Leakey, and Pickford disagree on the details: Michael Balter, "Paleontological Rift in the Rift Valley," *Science* 292 (2001), pp. 198–201.

114   Pickford says he took them out of the museum: Martin Pickford, interview in Nairobi, Kenya, September 15, 2003.

114   Pickford implicates Hill: Martin Pickford, interview in Nairobi, Kenya, September 15, 2003.

114   But he says he did not trust Pickford: Andrew Hill, telephone interview, April 3, 2005.

115   "I wasn't spying on him": Andrew Hill, telephone interview, April 3, 2005.

115   Richard Leakey told Senut that the ban: Brigitte Senut, interview in the Tugen Hills, Kenya, September 16, 2003; Richard Leakey, e-mail, June 17, 2005.

116   "I was banned from working in Kenya": Martin Pickford, interview in Nairobi, Kenya, September 15, 2003.

Part Two: The Decade of Discovery
117   "New and significant prehuman fossils have been unearthed": Stephen Jay Gould, "Our Greatest Evolutionary Step," in *The Panda's Thumb*, p. 125.

Chapter Eight: The Lady of the Lake
119   "The real voyage of discovery": Marcel Proust, *Remembrance of Things Past* (New York: Vintage Books, 1982), p. 260.

119   My main sources for the first section of this chapter are: interviews with Meave Leakey, en route to and at Kanapoi, Kenya, on October 20–31, 2003; interviews with Richard Leakey, a telephone interview on January 23, 2004, and an interview in Washington, D.C., on February 19, 2004; Meave Leakey, "The Dawn of Humans: The Farthest Horizon," *National Geographic,* September 1995, pp. 40–51; an interview with Alan Walker in Tampa, Florida, on April 14, 2004; Richard Leakey and Roger Lewin, *Origins Reconsidered* (New York: Doubleday, 1992), p. 24.

120   they often gazed across the lake and wondered: Richard Leakey and Roger Lewin, *Origins Reconsidered*, p. xix.

121   Kamoya convinced the young men: Alan Walker, interview in Tampa, Florida, April 14, 2004.

122   "We have something for you": Meave Leakey, "The Dawn of Humans," p. 40.

123   "Surely this is where we came from": "The Dawn of Humans," p. 40.

124   My main sources on Meave Leakey's background are: interviews with Meave Leakey, en route to and at Kanapoi, Kenya, on October 20–31, 2003; Virginia Morell, *Ancestral Passions.*

126   After five years of collecting, the team had only found: Meave G. Leakey and Alan C. Walker, "The Lothagam Hominids," in *Lothagam: The Dawn of Humanity in Eastern Africa*, ed. Meave Leakey and John M. Harris (New York: Columbia University Press, 2003), pp. 249–56.

127   On June 2, 1993, just a week after camp: Meave Leakey, interviews en route to and in Kanapoi, Kenya, October 20–31, 2003.

127   (Although some later suspected sabotage: Michael McRae, "Survival Test for Kenya's Wildlife," *Science* 280 (1998), p. 510.

127   Meave and a small scouting party: Meave Leakey, interviews en route to and in Kanapoi, Kenya, October 20–31, 2003.

127   Later, geologist Craig Feibel collected samples: Craig Feibel, telephone interview, December 11, 2003.

128   "Tim White has seventeen skeletons": Meave Leakey, interviews en route to and in Kanapoi, Kenya, October 20–31, 2003.

Chapter Nine: A View from Afar

129   "While I was reading about the Afar": Jon Kalb, *Adventures*, p. 17.

129   My main sources for the geology of the Afar are: Tim White, e-mail, November 10, 2004, and telephone interviews on October 29, 2001, and October 31, 2001; Tim White, in a CD of a talk at the 2003 Nobel Conference at Gustavus Adolphus College, October 7, 2003; Giday Wolde-Gabriel, interview in Florence, Italy, on August 22, 2004, and telephone interviews on July 1, 2003, and November 24, 2004; Yohannes Haile-Selassie, interviews in Cleveland, Ohio, and Florence, Italy, on March 10, 2004, and August 22, 2004; Paul Renne, telephone interview on February 8, 2005; Maurice Taieb, telephone interview on November 22, 2004.

Other sources included: "Oldest Human Ancestor Found in Ethiopia," Public Information Office, University of California at Berkeley, September 21, 2004; Tim D. White, Gen Suwa, and Berhane Asfaw, *"Australopithecus ramidus,* a New Species of Early Hominid from Aramis, Ethiopia," *Nature* 371 (1994), pp. 306–12; Giday WoldeGabriel et al., "Ecological and Temporal Placement of Early Pliocene Hominids at Aramis, Ethiopia," *Nature* 371 (1994), pp. 330–33; Giday WoldeGabriel et al., "Geoscience Methods Lead to Paleoanthropological Discoveries in Afar Rift, Ethiopia," *EOS, Transactions, American Geophysical Union,* July 20, 2004; Samson Tesfaye et al., "Early Continental Breakup Boundary and Migration of the Afar Triple Junction, Ethiopia," *GSA Bulletin* 115 (2003), pp. 1053–67.

129   even working with NASA scientists to learn: Ann Gibbons, "A New Look for Archeology," *Science* 252 (1991), pp. 918–20; Tim White, telephone interview, May 1, 1991.

130   three plates meet in a triple rift junction: Samson Tesfaye et al., "Early Continental Breakup"; Giday WoldeGabriel et al., "Volcanism, Tectonism, Sedimentation, and the Paleoanthropological Record in the Ethiopian Rift System," in *Volcanic Hazards and Disasters in Human Antiquity,* ed. F. W. McCoy and G. Heiken, *Geological Society of America Special Paper* 385 (2000), pp. 83–99.

130   has pushed the Afar triple junction about 160 kilometers: Samson Tesfaye et al., "Early Continental Breakup," p. 1053.

130   The engine for all this tectonic activity is a hot spot: Paul Renne, telephone interview, February 8, 2005; Giday WoldeGabriel, interview on August 22, 2004, in Florence, Italy; J. M. Kendall et al., "Magma-Assisted Rifting in Ethiopia," *Nature* 433 (2005), pp. 146–48.

133   WoldeGabriel and others take samples of the ancient soil: Giday WoldeGabriel, telephone interviews, July 1, 2003.

133   the bleached landscape is so blindingly bright: Tim White, lecture at the French Academy of Sciences in Paris, September 13, 2004.

134   Clark mailed Asfaw an application: Berhane Asfaw, telephone interview, June 27, 2003.

135   By the time he was invited: Giday WoldeGabriel, telephone interview, July 1, 2003.

136   He also approached the Ethiopian minister of culture: Berhane Asfaw, telephone interview, June 27, 2003.

Chapter Ten: The Root Ape

137   "The creature widely named as the Missing Link": Henry Gee, "Uprooting the Human Family Tree," *Nature* 373 (1995), p. 15.

137   On June 10, 1994, an editor: Henry Gee, e-mail, March 4, 2005; Henry Gee, *In Search of Deep Time: Beyond the Fossil Record to a New History of Life* (New York: Free Press, 1991), pp. 201–02; Berhane Asfaw, interview in Milwaukee, Wisconsin, April 7, 2005.

138   Gee, an associate editor: Henry Gee, e-mail, March 4, 2005.

138   He tantalized Gee with a set of monochrome: Tim White, e-mail, November 10, 2004; Henry Gee, *In Search of Deep Time*, p. 201.

139   it was alive 1.2 million years—or eighty thousand generations: "Oldest Human Ancestor Found in Ethiopia."

140   they got caught in a conflict: Giday WoldeGabriel, telephone interview, June 16, 2003.

140   As the midday sun bore down: Scott Simpson, interview in Cleveland, Ohio, November 21, 2004.

140   "I knew immediately that it was a hominid": Gen Suwa, quoted in "Oldest Human Ancestor Found in Ethiopia."

141   White would say later that the deciduous: Tim White, quoted in "Oldest Human Ancestor Found in Ethiopia."

144   "the most apelike hominid ancestor known": Tim White, et al., *"Australopithecus ramidus,* a New Species," p. 312.

144   "a long-sought potential root species": Tim White, et al., *"Australopithecus ramidus,* a New Species," p. 306.

145   "The metaphor of a 'missing link' ": Bernard Wood, "The Oldest Hominid Yet," *Nature* 371 (1994), p. 281.

145   "a single line and a single lineage": Tim White, quoted in Robert Lee Hotz, "Bones in Africa Called Closest Yet to Missing Link," *Los Angeles Times*, September 22, 1994, p. 1.

145   "oldest known link in the evolutionary chain": Tim White, quoted in "Oldest Human Ancestor Found in Ethiopia."

145   Henry Gee at *Nature* to eventually regret: Gee, *In Search of Deep Time*, p. 204.

146   Indeed, Pilbeam and Harvard primatologist Richard Wrangham: Richard Wrangham and David Pilbeam, "African Apes as Time Machines," in *All Apes Great and Small*, ed. B. M. F. Galdikas, N. E. Briggs, L. K. Sheeran, G. L. Shapiro, and J. Goodall, vol. 1: *Chimpanzees, Bonobos, and Gorillas* (New York: Kluwer Academic/Plenum, 2001), pp. 107–30.

149   He had warned that any identification: Elwyn Simons, "A View on the Science: Physical Anthropology at the Millennium," *American Journal of Physical Anthropology* 112 (2000), pp. 441–45.

149   "We have only begun to understand this ancestor": Tim White, quoted in "Oldest Human Ancestor Found in Ethiopia."

149   crawling along on hands and knees less than two hundred feet: Tim White, e-mail, November 10, 2004; Charles Petit, "Major Discovery of Prehuman Fossil," *San Francisco Chronicle*, January 15, 1995; John Noble Wilford, "Fossil Find May Show If Prehuman Walked," *New York Times*, February 21, 1995, p. B5.

150   White would exult: Tim White, e-mail, November 10, 2004.

151   when Meave Leakey came to visit him: Tim White, e-mail, June 26, 2005; Meave Leakey, interviews en route to and in Kanapoi, Kenya, October 20–31, 2003.

151   Leakey and Walker would soon make an announcement: Leakey et al., "New Four-Million-Year-Old Hominid Species," pp. 565–71.

152   "The next few years could see violent upsets": Henry Gee, "Uprooting the Human Family Tree."

Chapter Eleven: West Side Story

153   "The time had come": James Joyce, "The Dead," in *Dubliners* (New York: Viking Critical Library, 1969).

153   Yves Coppens was sitting at a table in an outdoor café: Yves Coppens, interview in Paris, France, October 5, 2002.

155   He would do the math: Yves Coppens, interview in Paris, France, October 5, 2002.

155   They had collected two thousand fossils of hominids: Yves Coppens, "East Side Story," p. 90.

155   Coppens was thinking about this puzzle: Yves Coppens, "East Side Story," p. 88.

156   Coppens called it the East Side Story: Yves Coppens, "East Side Story," p. 88.

157   "This is when I decided to change my topic": Michel Brunet, interview in Poitiers, October 1, 2002.

158   a lecture by Louis Leakey had influenced him: Michel Brunet, "Dreams of the Past," p. 121.

158   "What does he want?": Michel Brunet, interview in Poitiers, October 2, 2002.

158   "I am *Homo sapiens*, and I want to try to answer": Michel Brunet, e-mail, November 3, 2002.

159   in 1978, Brunet and his colleagues were standing on a flat rooftop: Michel Brunet, telephone interview, November 14, 2001.

159   Brunet and Heintz soon found themselves at the Silver Grill Hotel: Michel Brunet, telephone interview, April 2, 2005.

160   "We were two young guys," Michel Brunet, interview in Poitiers, October 3, 2002.

161   "We both decided Cameroon had passed its sell-by date": David Pilbeam, telephone interview, October 2002.

161   "Come on. You have to come to Chad!": David Pilbeam, telephone interview, October 2002.

161   two attempted coups in Chad in 1992 and 1993: "Background Note: Chad," U.S. Department of State, Bureau of African Affairs, February 2005, www.state.gov/r/pa/ei/bgn/37992.htm.

162   Beauvilain told Brunet it would be possible: Alain Beauvilain, telephone interview, December 2, 2004.

162   The French government paid Beauvilain's expenses: Beauvilain, telephone interview, December 2, 2004.

162–63 He would announce the discovery of an australopithecine: Michel Brunet, Alain Beauvilain, Yves Coppens, Emile Heintz, Aladji H. E. Moutaye, and David Pilbeam, "The First Australopithecine 2,500 Kilometres West of the Rift Valley (Chad)," *Nature* 378 (1995), pp. 273–75.

163   "The two windows that were open": Donald Johanson, telephone interview, January 26, 2005.

163   "We're not saying we know where the cradle": Marlise Simons, "New Species of Early Human Is Reported Found in Desert in Chad," *New York Times,* May 23, 1996.

164   two tents were destroyed by sandstorms: Alain Beauvilain, e-mail, December 2, 2004.

164   "The fraternity is more important": Michel Brunet, "Dreams of the Past," p. 121.

165   "Sometimes in the field you see something": Michel Brunet, interview in Poitiers, October 3, 2002.

165   Brunet had a heart attack: Michel Brunet, interview in Poitiers, October 1, 2002.

165   Brunet would express a sense of having too little time: Michel Brunet, "Dreams of the Past," p. 121.

Chapter Twelve: Turf Wars

166   "Under the unspoken rules of the paleoanthropological game": Donald Johanson, *Lucy,* p. 112.

167   "I do not approve": Martin Pickford, interview in Nairobi, Kenya, September 15, 2003.

167   dismiss Pickford as "evil": Meave Leakey, interview in Pittsburgh, October 17, 2002.

167   "Pickford truly enjoys screwing people": Andrew Hill, interview in Tampa, Florida, April 16, 2004.

167   the book said he had married his American wife: Eustace Gitonga and Martin Pickford, *Master of Deceit,* p. 126.

167   Pickford would say years later that he was not proud: Michael Balter, "Paleontological Rift."

167   "I only have one regret": Martin Pickford, interview in the Tugen Hills, Kenya, September 17, 2003.

168   comparing Richard Leakey to a giant boulder: Martin Pickford, field journal entry, "Kenya 1998–," September 18, 1999.

169   "You can see Mount Kilimanjaro": Eustace Gitonga, interview in Nairobi, Kenya, September 12, 2003.

169   Leakey and his "cartel": Eustace Gitonga, interview in Nairobi, Kenya, September 12, 2003.

169   Gitonga says he came up with the idea: Eustace Gitonga, interview in Nairobi, Kenya, September 12, 2003.

170   by 1995, only two black Kenyans: Martin Pickford, e-mail, September 10, 2005.

170   "scientists of good faith": Eustace Gitonga, "Discovery of Earliest Hominid Remains," *Science* 291 (2001), p. 986 (in "Letters").

171   an official in the department for issuing research permits: Letter to Martin Pickford, October 30, 1998, from J. E. Ekirapa in Office of the President, Kenya.

171   On January 4, 1999, Pickford met with Kiptoon: Martin Pickford, field journal entry, "Kenya 1998–", January 6, 1999.

171   "What's going on?" Feibel asked Hill: Andrew Hill, interview in New Haven, Connecticut, February 4, 2004.

172   George Abungu, demanded that the permit be revoked: Michael Balter, "Paleontological Rift."

172   "the letter is a crude and inept forgery": Martin Pickford, field journal entry, "Kenya 1998–," November 26, 1999.

172   Josephat Ekirapa, told *Science* in 2001: Michael Balter, "The Case of the 'Forged' Letter," *Science* 292 (2001), p. 200.

172   Pickford knew that the National Museums officials: Martin Pickford, field journal entry, August 20, 1999.

173   "I resent having to document how much time": Andrew Hill, interview in New Haven, Connecticut, February 4, 2004.

173   the project had been investigating the Tugen Hills continuously: Andrew Hill, "Paleoanthropological Research in the Tugen Hills, Kenya," *Journal of Human Evolution* 42 (2002), pp. 1–10.

174   Meave Leakey, Alan Walker, and their colleagues: Meave Leakey, Craig S. Feibel, Ian McDougall, Carol Ward, and Alan Walker, "New Specimens and Confirmation of an Early Age for *Australopithecus anamensis,*" *Nature* 393 (1998), pp. 62–66.

174   "at the moment, a Dr. Martin Pickford": Letter from Richard Leakey, head of Public Service, to George Abungu, director general, National Museums of Kenya, March 14, 2000.

175   sensing that it was some sort of trap: Martin Pickford, field journal entry, March 17, 2000.

175   mixing up their labels in the process of confiscating: Martin Pickford, field journal entry, March 17, 2000.

175 declaring it "nolle prosequi": Martin Pickford, "Discovery of Earliest Hominid Remains," *Science* 291 (2001), p. 986 (in "Letters").

176 "The fossils lose": Michael Balter, "Paleontological Rift," p. 200.

176 The main sources for the account of events at Sagatia are the author's interviews with Brigitte Senut at Sagatia, in the Tugen Hills, on September 17, 2003, and with Eustace Gitonga in Nairobi, Kenya, on September 19, 2003; the author's telephone interviews with Katie Binetti on June 22, 2005, and Andrew Hill on June 14, 2005; and letters from Eustace Gitonga on September 29, 2003, to Dr. Peter Salovey, dean of Yale Graduate School, to F. Clark Howell, University of California at Berkeley, and to Philip Rubin, division director, National Science Foundation.

176 she fumed that the Yale team: Brigitte Senut, interview at Sagatia, in the Tugen Hills, September 17, 2003.

176 "proceeded to raid a fossil site": Eustace Gitonga, in a letter to Peter Salovey, September 29, 2003.

177 He had a gun: Katie Binetti, telephone interview, June 22, 2005; Andrew Hill, telephone interview, June 14, 2005.

177 Gitonga told her he was there to deal with her: Katie Binetti, telephone interview, June 22, 2005.

177 Farah did: Idle O. Farah, e-mail, October 4, 2005.

179 Gitonga called it "liberalizing" multiple sectors: Eustace Gitonga, "Discovery of Earliest Hominid Remains," p. 986.

179 "undermining really good Kenyan laws": Michael Balter, "Paleontological Rift," p. 200.

179 Yellen at the NSF: Michael Balter, "Paleontological Rift," p. 200.

Chapter Thirteen: Toeing the Line

180 "Do not steal another person's site": Tim D. White, "A View on the Science," p. 291.

180 On April 13, 2000, an Ethiopian graduate: Yohannes Haile-Selassie, interview in San Antonio, Texas, on April 13, 2000.

181 and probably the first Ethiopian fossils of *Australopithecus anamensis:* Yohannes Haile-Selassie, "A Newly Discovered Early Pliocene Hominid Bearing Paleontological Site in the Mulu Basin, Ethiopia," Annual Meeting Issue, *American Journal of Physical Anthropology,* Suppl. 30 (2000), p. 170.

181   a team of foreign researchers had "claim-jumped" Galili: Yohannes Haile-Selassie, telephone interview, May 23, 2005; Yohannes Haile-Selassie, "Late Miocene Mammalian Fauna from the Middle Awash Valley, Ethiopia," Ph.D. diss., University of California, Berkeley, 2001.

181   This was not the first time an Ethiopian: Ann Gibbons, "Claim-Jumping Charges Ignite Controversy at Meeting," *Science* 268 (1995) pp. 196–97.

183   the Middle Awash Project members would suggest: Yohannes Haile-Selassie, telephone interview, May 23, 2005; Giday WoldeGabriel, telephone interview, June 13, 2004.

183   Seidler would protest vehemently: Horst Seidler, "Fossil Hunters in Dispute over Ethiopian Sites," *Nature* 411 (2001), p. 15 (in "Correspondence").

183   "On my word of honor, I never knew": Horst Seidler, telephone interview, June 6, 2005.

184   "completely inappropriate": Yohannes Haile-Selassie, telephone interview, May 23, 2005.

184   He included a photo: Yohannes Haile-Selassie, "Photos May Offer Clues over Ethiopian Fossil Site," *Nature* 412 (2001), p. 118 (in "Correspondence").

185   The field permits for White and the Middle Awash: Rex Dalton, "Restrictions Delay Fossil Hunts in Ethiopia," *Nature* 410 (2001), p. 728.

186   "In the field, do not think": Tim White, "A View on the Science."

187   Back in 1995, White and Asfaw had decided: Yohannes Haile-Selassie, telephone interview, May 23, 2005.

188   "We knew that the area was remote": Tim White, in "Earliest Hominids," press backgrounder prepared by the Middle Awash Research Group, June 30, 2001.

189   "How do you know it's a hominid?": Yohannes Haile-Selassie, interview in Cleveland, Ohio, March 10, 2004.

189   WoldeGabriel, who was showing White: Giday WoldeGabriel, telephone interview, November 24, 2004.

189   a reporter for the Cleveland *Plain Dealer:* John Mangels, "Fossil Hunter Transcripts."

191   he had identified the fossils as a new subspecies: Yohannes Haile-Selassie, "Late Miocene Hominids from the Middle Awash, Ethiopia," *Nature* 412 (2001), pp. 178–81.

192   "It was bad": Yohannes Haile-Selassie, telephone interview, May 23, 2005.

Chapter Fourteen: Millennium Man

193 "I have weathered the storm": Ezra Pound, *The Selected Poems of Ezra Pound* (New York: New Directions, 1957), p. 30.

193 "In this world there are two tragedies." Oscar Wilde, *Lady Windermere's Fan*, act 3, scene 1 (Mineola, New York: Dover Publications Inc., 1998), pp. 29–40.

193 On the morning of November 4, 2000: Martin Pickford, field journal entry, "Kenya 1998–," November 4, 2000.

195 the thighbone was found in four pieces: Martin Pickford, Brigitte Senut, Dominique Gommery, and Jacques Treil, "Bipedalism in *Orrorin tugenensis* Revealed by Its Femora," *Comptes Rendus Palevol* 1 (2002), p. 191; Tim White, "Early Hominid Femora: The Inside Story," *Comptes Rendus Palevol* (2005; in press).

195 "I find the complete skeleton of a small rat": Martin Pickford and Brigitte Senut in "The Secrets of the Dead," Washingtonpost.com (in Live Online), May 10, 2002.

195 "It's like reading a book": Martin Pickford, interview in the Tugen Hills, September 17, 2003.

196 "People would ruin us": Martin Pickford and Brigitte Senut in "The Secrets of the Dead," Washingtonpost.com (in Live Online), May 10, 2002.

196 "put Kenya on the map": "Random Samples, 6-Million-Year-Old Man," *Science* 290 (2000), p. 2065.

197 the thighbone was large and "human-like": Brigitte Senut, Martin Pickford, Dominique Gommery, Pierre Mein, Kiptalam Cheboi, and Yves Coppens, "First Hominid from the Miocene (Lukeino Formation, Kenya)," *Comptes Rendus de l'Académie des Sciences Paris, Sciences de la Terre et des Planètes* 332 (2001), p. 139.

197 The bandlike tendon puts pressure against the bone: Martin Pickford, Brigitte Senut, Dominique Gommery, and Jacques Treil, "Bipedalism in *Orrorin tugenensis,*" pp. 191–203.

198 "This simple phylogeny contrasts starkly": Leslie Aiello and Mark Collard, "Our Newest Oldest Ancestor?" *Nature* 410 (2001), p. 527.

199 "A year ago today": Martin Pickford, field journal entry, "Kenya 1998–," March 21, 2001.

199 Andrew Hill was at home, asleep: Andrew Hill, interview in New Haven, Connecticut, on February 4, 2004.

200 "not impressed with their evidence": John Noble Wilford, "On the Trail of a Few More Ancestors," *New York Times*, April 8, 2001, p. 6.

200   calling the groove on the thighbone the intertrochanteric groove: Tim D. White, "Early Hominid Femora."

200   "Given these irregularities": Tim D. White, "Early Hominid Femora."

201   upper thighbone for either species of *Ardipithecus:* Tim D. White, "Early Hominid Femora."

201   He also wanted to know *how* it walked upright: Ann Gibbons, "Oldest Femur Wades into Controversy," *Science* 305 (2004), p. 1885.

202   They also confirmed that the pattern of bone distribution: C. Owen Lovejoy, Richard S. Meindl, James C. Ohman, Kingsbury G. Heiple, and Tim D. White, "The Maka Femur and Its Bearing on the Antiquity of Human Walking: Applying Contemporary Concepts of Morphogenesis to the Human Fossil Record," *American Journal of Physical Anthropology* 119 (2002), p. 106.

202   like a cantilevered beam: C. Owen Lovejoy et al., "The Maka Femur," p. 111.

203   And the CT scans were of poor quality: James C. Ohman, C. Owen Lovejoy, and Tim D. White, "Questions About *Orrorin* Femur," *Science* 307 (2005), p. 845 (in "Letters").

204   there was no point: Martin Pickford, telephone interview, September 16, 2004.

204   the bone was broken in a zigzag pattern: Ann Gibbons, "Oldest Human Femur," p. 1885; Martin Pickford, telephone interview, September 16, 2004.

204   "Exceptional claims demand exceptional evidence": James C. Ohman, "Questions About *Orrorin* Femur," p. 845.

205   "I don't include *Orrorin*": Andrew Hill, e-mail, March 31, 2004.

205   "discoveries of fossil hominids are like buses": Henry Gee, "Return to the Planet of the Apes," *Nature* 412 (2001), p. 131.

206   "appears to be the most ancient human ancestor ever discovered": Michael D. Lemonick and Andrea Dorfman, "One Giant Step for Mankind," *Time* 158 (2001).

207   "Sadly, I doubt that the status of these creatures": Henry Gee, "Return to the Planet," p. 132.

Chapter Fifteen: Toumaï

208   "Is not the midnight like Central Africa": Henry David Thoreau, "Night and Moonlight," 1863, in vol. 5 of *The Writings of Henry David Thoreau* (Boston: Houghton Mifflin, 1906), p. 323.

208   "Ahounta, it is you who will find it": Ahounta Djimdoumalbaye, "The Discovery by Ahounta Djimdoumalbaye," Les Amis de la Paléontologie au Tchad Web site: www.chez.com/paleotchad.

210   "I was thus alone with the cranium": Ahounta Djimdoumalbaye, "The Discovery by Ahounta."

210   Fanoné Gongdibé, who had been trained: Fanoné Gongdibé, telephone interview, June 21, 2005.

210   Beauvilain thought it was a joke: Alain Beauvilain, *"Toumaï: L'Aventure Humaine* (Paris: La Table Ronde, 2003), p. 12.

211   "This is too big for us!": Alain Beauvilain, telephone interview, December 2, 2004.

211   "It is very moving to hold that skull in your hands": Fabrice Node-Langlois, "A New Ancestor of Man?" *Le Figaro,* July 31, 2001.

212   He was not at all pleased: Michel Brunet, e-mail, July 1, 2005.

212   They sat beside each other, mostly in silence: Alain Beauvilain, telephone interview, December 2, 2004.

213   he never had any intent of taking Brunet's place: Alain Beauvilain e-mail, June 20, 2005.

213   "It's a lot of emotion to have in my hand": Michel Brunet, in press release from *Nature,* July 10, 2002.

214   He would take Toumaï on a world tour: Michel Brunet, interview in Poitiers, France, October 1, 2002.

214   "As I looked it over": Dan Lieberman, telephone interview, July 4, 2002.

216   THE EARLIEST KNOWN HOMINID: Michel Brunet et al., "A New Hominid from the Upper Miocene of Chad, Central Africa," *Nature* 418 (2002), pp. 145–51; P. Vignaud et al., "Geology and Palaeontology of the Upper Miocene Toros-Menalla Hominid Locality, Chad," *Nature* 418 (2002), pp. 152–55.

217   "A single fossil can fundamentally change": Bernard Wood, "Hominid Revelations from Chad," *Nature* 418 (2002), pp. 133–34.

218   "*Sahelanthropus* or *Sahelpithecus?*": Milford Wolpoff, Brigitte Senut, Martin Pickford, and John Hawks, "*Sahelanthropus* or *Sahelpithecus?*" *Nature* 419 (2002), pp. 581–82 (in "Correspondence").

218   Brunet's reply in *Nature:* Michel Brunet, "Brunet et al. Reply," *Nature* 419 (2002), p. 582 (in "Correspondence").

219   "It is absolutely not a paleo-gorilla": Michel Brunet, interview in Poitiers, France, October 2, 2002.

219   "It's a tax": Michel Brunet, interview in Poitiers, France, October 3, 2002.

220   embarrassing reports of a "paternity battle": "Quarrel of Paternity Around the 'Oldest Human,' Toumaï," AFP News, September 18, 2003 (www.chez.com/paleotchad/texte/aafp18septembre2003.html).

220   Beauvilain thought that he and the other two Chadians: Alain Beauvilain, e-mail, June 19, 2005.

221   indeed, 45 percent of Americans who participated in a Gallup poll: Frank Newport, "Third of Americans Say Evidence Has Supported Darwin's Evolution Theory," Gallup News Service, November 19, 2004.

221   "It's part of what you have to do": Michel Brunet, interview in Poitiers, France, October 2, 2002.

Part Three: Wisdom of the Bones

223   "The question of questions for mankind": Thomas Henry Huxley, *Evidence as to Man's Place in Nature* (New York: D. Appleton, 1863).

Chapter Sixteen: Bones of Contention

226   a passionate debate among three scientists: Ann Gibbons, "Oldest Femur Wades into Controversy." The account of the debate is also based on the author's coverage of the meeting Prehistoric Climates, Cultures, and Societies at the French Academy of Sciences, September 13–14, 2004.

226   an article in the *South African Journal of Science:* Alain Beauvilain and Yves Le Guellec, "Further Details Concerning Fossils Attributed to *Sahelanthropus tchadensis* (Toumaï)," *South African Journal of Science* 100 (March–April 2004), pp. 142–44.

227   Pickford said is "absurd": Martin Pickford, e-mail, September 10, 2005.

227   Pickford, who was traveling: Martin Pickford, telephone interview, December 15, 2004.

227   He at first wrote a letter of apology: Martin Pickford, telephone interview, December 15, 2004.

228   "any competent morphologist:" David Pilbeam, e-mail, June 28, 2005.

228   Pickford and Beauvilain criticized the letter: Martin Pickford, telephone interview, December 15, 2004; Ann Gibbons, "Tooth Fight," *Science* 306 (2004), p. 2184.

228   Beauvilain wrote a new letter to the journal: Alain Beauvilain and Yves Le Guellec, "Beauvilain and Le Guellec Reply," *South African Journal of Science* 100 (September–October 2004), pp. 445–46 (in "Correspondence").

231   they would bring 144 fossils: This number comes from the author's count from published reports of the initial descriptions of the fossils of *Sahelanthropus tchadensis, Orrorin tugenensis, Ardipithecus kadabba, Ardipithecus ramidus,* and *Australopithecus anamensis* in *Nature, Science,* the *Journal of Human Evolution,* and *Comptes Rendus Palevol.*

231   a new partial skeleton of an upright-walking hominid: Ann Gibbons, "Skeleton of Upright Human Ancestor Discovered in Ethiopia," *Science* 307 (2005), pp. 1545–47.

231   "These dramatic discoveries" F. C. Howell and Tim White, "Revealing Hominid Origins Project Summary," Proposal to the National Science Foundation, p. 1.

232   the fossil record is exclusively African: Tim White, "Earliest Hominids," in *The Primate Fossil Record,* ed. Walter C. Hartwig (New York: Cambridge University Press, 2002), p. 407.

232   "it is somewhat more probable": Charles Darwin, *The Descent of Man.*

232   other than some isolated teeth: Sally McBrearty and Nina Jablonski, "First Fossil Chimpanzee," *Nature* 437 (2005), pp. 105–08, in press.

232   "You wouldn't invite *Ardipithecus ramidus*": Tim White, interview on National Public Radio's *Fresh Air* with Terry Gross, June 18, 2003.

233   This test, therefore, worked to identify hominids that were 4 million years: Tim White, "Earliest Hominids," p. 407.

233   an awkward period of transition: Henry M. McHenry, "Introduction to the Fossil Record of Human Ancestry," in *The Primate Fossil Record,* ed. Walter C. Hartwig (New York: Cambridge University Press, 2002), p. 404.

233   Toumaï meets the test: Tim White, e-mail, June 28, 2002.

234   "If we had skeletons": Owen Lovejoy, interview in Kent, Ohio, November 22, 2004.

234   paleoanthropologists have just scraped the surface: Bernard Wood, "Hominid Revelations," pp. 133–35.

235   "straight-ahead slog": Guy Gugliotta, "Earliest Human Ancestor?" *Washington Post*, July 11, 2002, p. 1.

235   "Since when is that evidence": Tim White, e-mail, June 26, 2005.

235   "Is the outlook completely gloomy?": Henry Gee, "Return to the Planet of the Apes," pp. 131–32.

Chapter Seventeen: Habitat for Humanity

236   "What is special is not the finding": Tim White, e-mail, November 10, 2004.

236   "You can know the name of a bird": Richard Feynman, "What Is Science?" paper delivered at fifteenth annual meeting of the National Science Teachers Association, New York City, 1966, and printed in *Physics Teacher* 7 (1968), pp. 313–20.

236   Michel Brunet was animatedly describing the type of environment: Michel Brunet and Richard Wrangham, interview at the Casino in Chicago, October 11, 2003.

237   He began a recent article on the origins of upright walking: Henry M. McHenry, "Origin of Human Bipedality," *Evolutionary Anthropology* 13 (2004), pp. 116–19.

238   cooling over the past 50 million years: Elisabeth S. Vrba, George H. Denton, Timothy C. Partridge, and Lloyd H. Burckle, eds., *Paleoclimate and Evolution with Emphasis on Human Origins* (New Haven: Yale University Press, 1995), pp. 9, 54.

239   woodland gave way to more grasslands: Thure Cerling, J. M. Harris, B. J. MacFadden, M. G. Leakey, and others, "Global Vegetation Change Through the Miocene/Pliocene Boundary," *Nature* 389 (1997), pp. 153–58.

239   some mammals were migrating: Yohannes Haile-Selassie, "Late Miocene Mammalian Fauna from the Middle Awash Valley, Ethiopia," p. 366; Giday WoldeGabriel, Yohannes Haile-Selassie, Paul R. Renne, William K. Hart, Stanley H. Ambrose, Berhane Asfaw, Grant Heiken, and Tim White, "Geology and Palaeontology of the Late Miocene Middle Awash, Afar Rift, Ethiopia," *Nature* 412 (2001), pp. 175–78.

240   once covered an area that was 400,000 square kilometers: Michel Brunet, in a talk to the Regional Council of Poitou-Charentes, October 2, 2002.

240   It was the kind of landscape: Jean-Renaud Boisserie, report at the thirty-second International Geological Congress in Florence, Italy, August 21, 2004; P. Vignaud et al., "Geology and Palaeontology," pp. 152–55.

240   On April 7, 2005, Toumaï was back on the cover: Michel Brunet et al., "New Material of the Earliest Hominid from the Upper Miocene of Chad," *Nature* 434 (2005), pp. 752–55; Christophe

P. E. Zollikofer et al., "Virtual Cranial Reconstruction of *Sahelanthropus tchadensis,* " *Nature* 434 (2005), pp. 755–59.

241   they both ended up giving Toumaï a face-lift: Ann Gibbons. "Facelift Supports Skull's Status as the Oldest Member of the Human Family," *Science* 308 (2005), pp. 179–81.

242   they could not get it to fit the shape of a gorilla: Christophe Zollikofer and Marcia Ponce de León, telephone interview, March 30, 2005.

242   "You need postcranial fossils to be one hundred percent": Dan Lieberman, telephone interview, March 11, 2005.

242   "I consider that he's wading across the same swamp": Martin Pickford, e-mail, October 15, 2004.

243   Brunet left Qaddafi's tent: Michel Brunet, telephone interview, April 2, 2005.

243   "More will be coming," Michel Brunet, telephone interview, April 2, 2005; Ann Gibbons, "Facelift," p. 181.

BIBLIOGRAPHY

Aiello, Leslie, and Mark Collard. "Our Newest Oldest Ancestor?" *Nature* 410 (2001), p. 527.

Andrews, Peter. *"Ramapithecus wickeri* Mandible from Fort Ternan, Kenya." *Nature* 230 (1971), pp. 192–94.

Andrews, Roy Chapman. *Meet Your Ancestors*. New York: Viking Press, 1945.

Balter, Michael. "The Case of the 'Forged' Letter." *Science* 292 (2001), p. 200.

———. "Paleontological Rift in the Rift Valley." *Science* 292 (2001), pp. 198–201.

Beauvilain, Alain. *Toumaï: L'Aventure Humaine*. Paris: La Table Ronde, 2003.

Beauvilain, Alain, and Yves Le Guellec. "Further Details Concerning Fossils Attributed to *Sahelanthropus tchadensis* (Toumaï)." *South African Journal of Science* 100 (March–April 2004), pp. 142–44.

———. "Beauvilain and Le Guellec Reply." *South African Journal of Science* 100 (September–October 2004), pp. 445–46 (in "Correspondence").

Bell, Robert. *Impure Science: Fraud, Compromise, and Political Influence in Scientific Research*. New York: John Wiley & Sons, 1992.

Brockman, John, ed. *Curious Minds: How a Child Becomes a Scientist*. New York: Pantheon Books, 2004.

Broom, Robert. *Finding the Missing Link*. London: Watts and Co., 1950.

Brunet, Michel. In "Dreams of the Past," *Nature* 423 (2003), p. 121.

Brunet, Michel, Alain Beauvilain, Yves Coppens, Emile Heintz, Aladji H. E. Moutaye, and David Pilbeam. "The First Australopithecine 2,500 Kilometres West of the Rift Valley (Chad)." *Nature* 378 (1995), pp. 273–75.

Brunet, Michel, F. Guy, D. Pilbeam, H. Mackaye, A. Likius, D. Ahounta, A. Beauvilain, C. Blondel, H. Bocherens, J. R. Boisserie, L. De Bonis, Y. Coppens, J. Deja, C. Denys, P. Duringer, V. Eisenmann, G. Fanone, P. Fronty, D. Geraads, T. Lehmann, F. Lihoreau, A. Louchar, A. Mahamat, G. Merceron, G. Mouchelin, O. Otero, P. Campomanes, M. Ponce de León, J. C. Rage, M. Sapanet, M. Schuster, J. Sudre, P. Tassy, X. Valentin, P. Vignaud, L. Viriot, A. Zazzo, and C. Zollikofer. "A New Hominid from the Upper Miocene of Chad, Central Africa." *Nature* 418 (2002), pp. 145–51.

———. "Brunet et al. Reply." *Nature* 419 (2002), p. 582 (in "Correspondence").

Brunet, Michel, and MPFT. *"Sahelanthropus tchadensis:* The Facts." *South African Journal of Science* 100 (September–October 2004), pp. 443–44 (in "Correspondence").

Brunet, Michel, et al. "New Material of the Earliest Hominid from the Upper Miocene of Chad." *Nature* 434 (2005), pp. 752–55.

Campbell, Bernard. *Human Evolution: An Introduction to Man's Adaptations.* Chicago: Aldine Publishing Co., 1966.

Cerling, Thure, J. M. Harris, B. J. MacFadden, M. G. Leakey, and others. "Global Vegetation Change Through the Miocene/Pliocene boundary." *Nature* 389 (1997), pp. 153–58.

Coppens, Yves. "East Side Story: The Origin of Humankind." *Scientific American,* May 1994, pp. 88–95.

———. *Lucy's Knee.* Pretoria: Protea Book House, 2002.

Dalton, Rex. "Restrictions Delay Fossil Hunts in Ethiopia." *Nature* 410 (2001), p. 728.

———. "Brickbats for Fossil Hunter Who Claims Skull Has False Tooth." *Nature* 430 (2004), p. 956.

Dart, Raymond A., with Dennis Craig. *Adventures with the Missing Link.* New York: Harper & Brothers, 1959.

Darwin, Charles. *On the Origin of Species by Means of Natural Selection.* London: John Murray, 1859; repr. Middlesex: Penguin Books, 1968.

———. *The Descent of Man and Selection in Relation to Sex.* London: John Murray, 1871; repr. London: Penguin Books, 2004.

————. *Diary of the Voyage of H.M.S.* Beagle. In vol. 1 of *The Works of Charles Darwin*. New York: New York University Press, 1987.

Dobzhansky, Theodosius. *Mankind Evolving: The Evolution of the Human Species*. New Haven: Yale University Press, 1962.

Fellows, Lawrence. "Man's Ancestor Found to Be More Than 19 Million Years Old." *New York Times*, January 15, 1967.

Feynman, Richard. "What Is Science?" Paper delivered at the fifteenth annual meeting of the National Science Teachers Association, New York City, 1966. Printed in *Physics Teacher* 7 (1968), pp. 313–20.

Gee, Henry. *In Search of Deep Time: Beyond the Fossil Record to a New History of Life*. New York: Free Press, 1991.

————. "Uprooting the Human Family Tree." *Nature* 373 (1995), p. 15.

————. "Return to the Planet of the Apes." *Nature* 412 (2001), p. 132.

Gibbons, Ann. "Anthropology Goes Back to Ethiopia." *Science* 249 (1990), p. 1373.

————. "A New Look for Archeology." *Science* 252 (1991), pp. 918–20.

————. "When It Comes to Evolution, Humans Are in the Slow Class." *Science* 267 (1995), pp. 1907–08.

————. "Claim-Jumping Charges Ignite Controversy at Meeting." *Science* 268 (1995), pp. 196–97.

————. "What Genes Make Us Human?" *Science* 281 (1998), pp. 1432–34.

————. "In Search of the First Hominids." *Science* 295 (2002), pp. 1214–18.

————. "First Member of the Human Family Uncovered." *Science* 297 (2002), pp. 171–73.

————. "Glasnost for Hominids: Seeking Access to Fossils." *Science* 297 (2002), pp. 1464–68.

————. "Africans Begin to Make Their Mark in Human Origins Research." *Science* 301 (2003), pp. 1178–79.

————. "Oldest Femur Wades into Controversy." *Science* 305 (2004), pp. 1885.

————. "Tooth Fight." *Science* 306 (2004), p. 2184.

————. "Skeleton of Upright Human Ancestor Discovered in Ethiopia." *Science* 307 (2005), pp. 1545–47.

————. "Facelift Supports Skull's Status as the Oldest Member of the Human Family." *Science* 308 (2005), pp. 179–81.

Gitonga, Eustace. "Discovery of Earliest Hominid Remains." *Science* 291 (2001), p. 986 (in "Letters").

Gitonga, Eustace, and Martin Pickford. *Richard E. Leakey: Master of Deceit.* Nairobi: White Elephant Publishers, 1995.

Glen, William. *The Road to Jaramillo: Critical Years of the Revolution in Earth Science.* Stanford: Stanford University Press, 1982.

Goodman, Morris. "Epilogue: A Personal Account of the Origins of a New Paradigm." *Molecular Phylogenetics and Evolution* 5 (1996), pp. 269–85.

Gould, Stephen Jay. *Ontogeny and Phylogeny.* Cambridge: Belknap Press–Harvard University Press, 1977.

————. *The Panda's Thumb.* New York: W. W. Norton, 1980.

————. *The Mismeasure of Man.* New York: W. W. Norton, 1981.

Gugliotta, Guy. "Earliest Human Ancestor?" *Washington Post,* July 11, 2002, p. 1.

Haeckel, Ernst. *The History of Creation, or The Development of the Earth and Its Inhabitants by the Action of Natural Causes: A Popular Exposition of the Doctrine of Evolution in General, and That of Darwin, Goethe, and Lamarck in Particular.* Translated by E. Ray Lankester. New York: D. Appleton, 1868.

Haile-Selassie, Yohannes. "A Newly Discovered Early Pliocene Hominid Bearing Paleontological Site in the Mulu Basin, Ethiopia." Annual Meeting Issue, *American Journal of Physical Anthropology,* Suppl. 30 (2000), p. 170.

————. "Photos May Offer Clues over Ethiopian Fossil Site." *Nature* 412 (2001), p. 118 (in "Correspondence").

————. "Late Miocene Mammalian Fauna from the Middle Awash Valley, Ethiopia." Ph.D. diss., University of California, Berkeley, 2001.

————. "Late Miocene Hominids from the Middle Awash, Ethiopia." *Nature* 412 (2001), p. 178–81.

Haile-Selassie, Yohannes, G. Suwa, and T. D. White. "Late Miocene Teeth from Middle Awash, Ethiopia, and Early Hominid Dental Evolution." *Science* 303 (2004), pp. 1503–05.

Hartwig, Walter Carl, ed. *The Primate Fossil Record*. New York: Cambridge University Press, 2002.

Hill, Andrew. "Early Hominid from Baringo, Kenya." *Nature* 315 (1985), pp. 222–24.

———. "Paleoanthropological Research in the Tugen Hills, Kenya." *Journal of Human Evolution* 42 (2002), pp. 1–10.

Hotz, Robert Lee. "Bones in Africa Called Closest Yet to Missing Link," *Los Angeles Times*, September 22, 1994, p. 1.

Howell, F. Clark, and the Editors of Life. *Early Man*. New York: Time, 1965.

Huxley, Thomas Henry. *Evidence as to Man's Place in Nature*. New York: D. Appleton, 1863.

Johanson, Donald, and Maitland Edey. *Lucy: The Beginnings of Humankind*. New York: Simon & Schuster, 1981.

Johanson, Donald, and James Shreeve. *Lucy's Child*. New York: William Morrow, 1989.

Jones, Steve, Robert Martin, and David Pilbeam, eds. *The Cambridge Encyclopedia of Human Evolution*. Cambridge: Cambridge University Press, 1992.

Joyce, James. "The Dead." In *Dubliners*. New York: Viking Critical Library, 1969.

Kalb, Jon. *Adventures in the Bone Trade*. New York: Copernicus Books, 2001.

Kelly, Jay. "Sexual Dimorphism in Canine Shape Among Extant Great Apes." *American Journal of Physical Anthropology* 96 (1995), pp. 365–89.

———. "Phylogeny and Sexually Dimorphic Characters: Canine Reduction in Ouranopithecus." In *Hominoid Evolution and Climatic Change in Europe*. Vol. 2 of *Phylogeny of the Neogene Hominoid Primates of Eurasia*. Edited by L. de Bonis, G. D. Koufos, and Peter Andrews. Cambridge: Cambridge University Press, 2001.

Kendall, J. M., G. W. Stuart, C. J. Ebinger, I. D. Bastow, and D. Keir. "Magma-Assisted Rifting in Ethiopia." *Nature* 433 (2005), pp. 146–48.

Kingdon, Jonathan. *Lowly Origin: Where, When and Why Our Ancestors First Stood Up*. Princeton: Princeton University Press, 2003.

Klein, Richard. *The Human Career: Human Biological and Cultural Origins.* Chicago: University of Chicago Press, 1999.

Kreisler, Harry. "On the Trail of Our Human Ancestors." Interview with Tim D. White on September 18, 2003. Conversations with History. Institute of International Studies, University of California, Berkeley. Published on the Internet at http://globetrotter.berkeley.edu/people3/White/white-con1.html.

Leakey, L. S. B. "Finding the World's Earliest Man." *National Geographic,* September 1960, pp. 420–35.

———. *White African.* New York: Ballantine Books, 1966.

———. "The Relationship of African Apes, Man, and Old World Monkeys." *Proceedings of the National Academy of Sciences* 67 (1970), pp. 746–48.

———. *By the Evidence: Memoirs, 1932–1951.* New York: Harcourt Brace Jovanovich, 1974.

Leakey, Mary. *Disclosing the Past: An Autobiography.* New York: Doubleday, 1984.

Leakey, Meave. "The Dawn of Humans: the Farthest Horizon." *National Geographic,* September 1995.

Leakey, Meave, Craig S. Feibel, Ian McDougall, and Alan Walker. "New Four-Million-Year-Old Hominid Species from Kanapoi and Allia Bay, Kenya." *Nature* 375 (1995), pp. 565–71.

Leakey, Meave, Craig S. Feibel, Ian McDougall, Carol Ward, and Alan Walker. "New Specimens and Confirmation of an Early Age for *Australopithecus anamensis."* *Nature* 393 (1998), pp. 62–66

Leakey, Meave, and John M. Harris, eds. *Lothagam: The Dawn of Humanity in Eastern Africa.* New York: Columbia University Press, 2003.

Leakey, Meave, and Alan Walker. "Early Hominid Fossils from Africa." *Scientific American,* Special Edition, August 25, 2003, pp. 14–19.

Leakey, Richard, and Roger Lewin. *Origins Reconsidered.* New York: Doubleday, 1992.

Lemonick, Michael D., and Andrea Dorfman. "One Giant Step for Mankind," *Time* 158 (2001).

Lewin, Roger. "Ethiopia Halts Prehistory Research." *Science* 219 (1983), pp. 147–49.

———. *Bones of Contention.* New York: Simon & Schuster, 1987.

———. *Patterns in Evolution: The New Molecular View.* New York: Scientific American Library, 1999.

————. "The Old Man of Olduvai Gorge." *Smithsonian*, October 2002.

Lovejoy, C. Owen, Richard S. Meindl, James C. Ohman, Kingsbury G. Heiple, and Tim D. White. "The Maka Femur and Its Bearing on the Antiquity of Human Walking: Applying Contemporary Concepts of Morphogenesis to the Human Fossil Record." *American Journal of Physical Anthropology* 119 (2002), p. 106.

Lovejoy, Owen. "Proximal Femoral Anatomy of *Australopithecus.*" *Nature* 235 (1972), pp. 175–76.

MacLatchy, Laura. "The Oldest Ape." *Evolutionary Anthropology* 13.3 (2004), pp. 90–103.

Mangels, John. "Fossil Hunter Transcripts." Cleveland *Plain Dealer*, November 28, 2004.

Marks, Jonathan. "Sherwood Washburn, 1911–2000." *Evolutionary Anthropology* 9.6 (2001), pp. 225–26.

Marshall, Eliot. "Gossip and Peer Review at the NSF." *Science* 238 (1987), p. 1502.

Mayr, Ernst. *The Growth of Biological Thought: Diversity, Evolution, and Inheritance.* Cambridge: Belknap Press–Harvard University Press, 1982.

————. *What Evolution Is.* New York: Perseus Books Group, 2001.

McBrearty, Sally, and Nina Jablonski. "First Fossil Chimpanzee." *Nature* 437 (2005), pp. 105–08.

McHenry, Henry M. "Introduction to the Fossil Record of Human Ancestry." In *The Primate Fossil Record*, edited by Walter C. Hartwig. Cambridge: Cambridge University Press, 2002, p. 405.

————. "Origin of Human Bipedality." *Evolutionary Anthropology* 13 (2004), pp. 116–19.

McRae, Michael. "Survival Test for Kenya's Wildlife." *Science* 280 (1998), p. 510.

Morell, Virginia. *Ancestral Passions: The Leakey Family and the Quest for Humankind's Beginnings.* New York: Simon & Schuster, 1995.

Newport, Frank. "Third of Americans Say Evidence Has Supported Darwin's Evolution Theory," Gallup News Service, November 19, 2004 (http://poll.gallup.com).

Ogot, Bethwell A. *My Footprints on the Sands of Time.* Canada: Trafford Publishers, 2003.

Ohman, James C., C. Owen Lovejoy, and Tim D. White. "Questions About *Orrorin* Femur." *Science* 307 (2005), p. 845 (in "Letters").

"Oldest Human Ancestor Found in Ethiopia," Public Information Office, University of California at Berkeley, September 21, 2004.

Patterson, Bryan, Anna K. Behrensmeyer, and William D. Sill. "Geology and Fauna of a New Pliocene Locality in North-western Kenya." *Nature* 266 (1970), pp. 918–21.

Petit, Charles. "Major Discovery of Prehuman Fossil." *San Francisco Chronicle*, January 15, 1995.

Pickford, Martin. *Louis S. B. Leakey: Beyond the Evidence*. London: Janus Publishing, 1997.

———. "Discovery of Earliest Hominid Remains." *Science* 291 (2001), p. 986 (in "Letters").

Pickford, Martin, and Brigitte Senut. "The Geological and Faunal Context of Late Miocene Hominid Remains from Lukeino, Kenya." *Comptes Rendus de l'Académie des Sciences* Ser. IIa 332 (2001), pp. 145–52.

———. "Hominoid Teeth with Chimpanzee- and Gorilla-like Features from the Miocene of Kenya: Implications for the Chronology of the Ape-Human Divergence and Biogeography of Miocene Hominoids." *Anthropological Science* 113 (2005), pp. 95–102.

Pickford, M., D.C. Johanson, C. O. Lovejoy, T. D. White, and J. L. Aronson. "A Hominoid Humeral Fragment from the Pliocene of Kenya." *American Journal of Physical Anthropology* 60 (1983), pp. 337–46.

Pickford, Martin, Brigitte Senut, Dominique Gommery, and Jacques Treil. "Bipedalism in *Orrorin tugenensis* Revealed by Its Femora." *Comptes Rendus Palevol* 1 (2002), pp. 191–203.

Pilbeam, David. "Notes on *Ramapithecus,* the Earliest Known Hominid, and *Dryopithecus.*" *American Journal of Physical Anthropology* 25 (1966), pp. 1–5.

———. "The Descent of Hominoids and Hominids." *Scientific American* (March 1984), pp. 84–96.

———. "Patterns of Hominoid Evolution." In *Ancestors: The Hard Evidence*, edited by Eric Delson. New York: Alan R. Liss, 1985.

———. "Distinguished Lecture: Hominoid Evolution and Hominoid Origins." *American Anthropologist* 88 (1986), pp. 295–312.

———. "Perspectives on the Miocene Hominoidea." In *The Primate Fossil Record*, edited by Walter Hartwig. New York: Cambridge University Press, 2002.

Pilbeam, David, and Nathan Young. "Hominoid Evolution: Synthesizing Disparate Data." *Comptes Rendus Palevol* 3 (2004), pp. 305–21.

Pound, Ezra. *The Selected Poems of Ezra Pound*. New York: New Directions, 1957.

Proust, Marcel. *Remembrance of Things Past*. New York: Vintage Books, 1982.

Reader, John. *Missing Links: The Hunt for Earliest Man.* Boston: Little, Brown, 1981.

Sarich, Vincent. "Immunological Evidence on Primates." In *The Cambridge Encyclopedia of Human Evolution.* Edited by Steve Jones, Robert Martin, and David Pilbeam. Cambridge: Cambridge University Press, 1992, pp. 303–06.

Sarich, Vincent, and P. J. Dolhinow. "A Molecular Approach to the Question of Human Origins." In *Background for Man.* Boston: Little, Brown, 1971, pp. 60–81.

Sarich, Vincent, and A. C. Wilson. "Immunological Time Scale for Hominid Evolution." *Science* 158 (1967), pp. 1200–03.

Seidler, Horst. "Fossil Hunters in Dispute over Ethiopian Sites." *Nature* 411 (2001), p. 15 (in "Correspondence").

Selvin, Paul. "The Raging Bull of Berkeley," *Science* 251 (1991), p. 368.

Senut, Brigitte, Martin Pickford, Dominique Gommery, Pierre Mein, Kiptalam Cheboi, and Yves Coppens. "First Hominid from the Miocene (Lukeino Formation, Kenya)." *Comptes Rendus de l'Académie des Sciences Paris, Sciences de la Terre et des Planètes* 332 (2001), p. 139.

Senut, Brigitte, and Martin Pickford. "La Dichotomie Grands Singes-Homme Revisitée." *Comptes Rendus Palevol* 3 (2004), pp. 265–76.

Sexton, Anne. "For Mr. Death Who Stands with His Door Open." *The Death Notebooks.* Boston: Houghton Mifflin, 1974.

Shipman, Pat. *The Man Who Found the Missing Link.* New York: Simon & Schuster, 2001.

Simons, Elwyn. "The Origin and Radiation of the Primates." *Annals of the New York Academy of Sciences* 167 (1968), pp. 319–31.

———. "Human Origins." *Science* 245 (1987), pp. 1343–50.

———. "A View on the Science: Physical Anthropology at the Millennium." *American Journal of Physical Anthropology* 112 (2000), pp. 441–45.

Simons, Elwyn, and David Pilbeam. "Preliminary Revision of Dryopithecinae (Pongidae, Anthropoidea)." *Folia Primatologica* (1965), pp. 81–152.

Simons, Marlise. "New Species of Early Human Is Reported Found in Desert in Chad." *New York Times,* May 23, 1996.

Stanford, Craig. *Upright: The Evolutionary Key to Becoming Human.* Boston: Houghton Mifflin, 2003.

Steiper, Michael, Nathan M. Young, and Tika Y. Sukarna. "Genomic Data Support the Hominoid-Cercopithecoid Divergence." *Proceedings of the National Academy of Sciences,* November 9, 2004, pp. 17021–26.

Strier, Karen B. *Primate Behavioral Ecology.* Boston: Allyn and Bacon, 2000.

Sullivan, Walter. "Bone Found in Kenya Indicates Man Is 2.5 Million Years Old." *New York Times,* January 14, 1967.

Swisher, Carl III, G. H. Curtis, T. Jacob, A. G. Getty, and A. Suprijo Widlasmoro. "Age of the Earliest Known Hominids in Java, Indonesia." *Science* 263 (1994), pp. 1118–21.

Tattersall, Ian. *The Fossil Trail.* New York: Oxford University Press, 1995.

Terradillos, Jean-Luc. "Abel: L'Homme de la Rivière aux Gazelles." *L'Actualité Poitou-Charentes* 31 (1996), pp. 16–21.

Tesfaye, Samson, et al. "Early Continental Breakup Boundary and Migration of the Afar Triple Junction, Ethiopia." *GSA Bulletin* 115 (2003), pp. 1053–67.

Thoreau, Henry David. "Night and Moonlight." 1863. In vol. 5 of *The Writings of Henry David Thoreau.* Boston: Houghton Mifflin, 1906, p. 323.

Turnbull, William D., and Farish A. Jenkins Jr. "Bryan Patterson, 1901–1979." *Society of Vertebrate Paleontology News Bulletin* February 1981.

van der Made, Jan. "Methods in Biostratigraphy: A Reply to Pickford et al." *Geobios* 36 (2003), pp. 223–28.

Vignaud, P., P. Duringer, H. T. Mackaye, A. Likius, C. Blondel, J. R. Boisserie, L. de Bonis, V. Eisenman, M. E. Étienne, D. Geraads, F. Guy, T. Lehmann, F. Lihoreau, N. Lopez-Martinez, C. Mourer-Chauviré, O. Otero, J. C. Rage, M. Schuster, L. Viriot, A. Zazzo, and M. Brunet. "Geology and Palaeontology of the Upper Miocene Toros-Menalla Hominid Locality, Chad." *Nature* 418 (2002), pp. 152–55.

Vrba, Elisabeth S., George H. Denton, Timothy C. Partridge, and Lloyd H. Burckle, eds. *Paleoclimate and Evolution, with Emphasis on Human Origins.* New Haven: Yale University Press, 1995.

Walker, Alan, and Pat Shipman. *The Ape in the Tree: An Intellectual and Natural History of* Proconsul. Cambridge: Belknap Press–Harvard University Press, 2005.

———. *The Wisdom of the Bones.* New York: Random House, 1996.

Ward, Steve, Barbara Brown, Andrew Hill, Jay Kelley, and Will Downs. "Equatorius: A New Hominoid Genus from the Middle Miocene of Kenya." *Science* 27, August 1999.

Ward, Steve, and Dana Duren. "Middle and Late Miocene African Hominoids." In *The Primate Fossil Record*, edited by Walter Hartwig. New York: Cambridge University Press, 2002.

Washburn, S. L., ed. *Classification and Human Evolution*. Chicago: Aldine, 1963.

Western, David. "Wildlife Conservation in Kenya." *Science* 280 (1998); p. 1507 (in "Letters").

White, Tim D. "A View on the Science: Physical Anthropology at the Millennium." *American Journal of Physical Anthropology* 112 (2000), pp. 287–92.

————. "Earliest Hominids." In *The Primate Fossil Record*, edited by Walter C. Hartwig. New York: Cambridge University Press, 2002, p. 407.

————. "Early Hominid Femora: The Inside Story." *Comptes Rendus Palevol* (2005), in press.

White, Tim D., Gen Suwa, and Berhane Asfaw. "*Australopithecus ramidus*, a New Species of Early Hominid from Aramis, Ethiopia." *Nature* 371 (1994), pp. 306–12.

————. "Corrigendum: *Australopithecus ramidus*, a New Species of Early Hominid from Aramis, Ethiopia." *Nature* 375 (1995), p. 88.

Wilde, Oscar. *Lady Windermere's Fan*. Mineola, New York: Dover Publications Inc., 1998.

Wilford, John Noble. "Fossil Find May Show If Prehuman Walked." *New York Times*, February 21, 1995, p. B5.

———— "In Kenya Fossils Are Found of Earliest Walking Species." *New York Times*, August 17, 1995, p. A1.

————. "On the Trail of a Few More Ancestors." *New York Times*, April 8, 2001, p. 6.

Willis, Delta. *The Hominid Gang*. New York: Viking Penguin, 1989.

WoldeGabriel, G., G. Heiken, T. D. White, B. Asfaw, W. K. Hart, and P. R. Renne. "Volcanism, Tectonism, Sedimentation, and the Paleoanthropological Record in the Ethiopian Rift System." In *Volcanic Hazards and Disasters in Human Antiquity*, edited by F. W. McCoy and G. Heiken. *Geological Society of America Special Paper* 385 (2000), pp. 83–99.

WoldeGabriel, Giday, T. D. White, G. Suwa, P. Renne, J. de Heinzelin, W. K. Hart, and G. Helken. "Ecological and Temporal Placement of Early Pliocene Hominids at Aramis, Ethiopia." *Nature* 371 (1994), pp. 330–33.

WoldeGabriel, Giday, Yohannes Haile-Selassie, Paul R. Renne, William K. Hart, Stanley H. Ambrose, Berhane Asfaw, Grant Heiken, and Tim White. "Geology and Palaeontology of the Late Miocene Middle Awash, Afar Rift, Ethiopia." *Nature* 412 (2001), pp. 175–78.

WoldeGabriel, et al. "Geoscience Methods Lead to Paleoanthropological Discoveries in Afar Rift, Ethiopia." *EOS, Transactions, American Geophysical Union,* July 20, 2004.

Wolpoff, Milford, Brigitte Senut, Martin Pickford, and John Hawks. *"Sahelanthropus* or *Sahelpithecus?" Nature* 419 (2002), pp. 581–82 (in "Correspondence").

Wood, Bernard. "The Oldest Hominid Yet." *Nature* 371 (1994), pp. 280–81.

———. "Hominid Revelations from Chad." *Nature* 418 (2002), pp. 133–34.

Wrangham, Richard, and David Pilbeam. "African Apes as Time Machines." In *All Apes Great and Small,* edited by B. M. F. Galdikas, N. E. Briggs, L. K. Sheeran, G. L. Shapiro, and J. Goodall. Volume 1: *Chimpanzees, Bonobos, and Gorillas.* New York: Kluwer/Plenum, 2001.

Zollikofer, Christophe P. E. et al., "Virtual Cranial Reconstruction of *Sahelanthropus tchadensis." Nature* 434 (2005), pp. 755–59.

INDEX

About the Author

ANN GIBBONS, the primary writer on human evolution for *Science* magazine for more than a decade, has taught science writing at Carnegie Mellon University. She lives in Pittsburgh, Pennsylvania.